Luca Alloatti

High-Speed, Low-Power and Mid-IR Silicon Photonics Applications

Karlsruhe Series in Photonics & Communications, Vol. 12
Edited by Profs. J. Leuthold, W. Freude and C. Koos

Karlsruhe Institute of Technology (KIT)
Institute of Photonics and Quantum Electronics (IPQ)
Germany

High-Speed, Low-Power and Mid-IR Silicon Photonics Applications

by
Luca Alloatti

KIT Scientific Publishing

Dissertation, Karlsruher Institut für Technologie (KIT)
Fakultät für Elektrotechnik und Informationstechnik, 2012

Impressum

Karlsruher Institut für Technologie (KIT)
KIT Scientific Publishing
Straße am Forum 2
D-76131 Karlsruhe
www.ksp.kit.edu

KIT – Universität des Landes Baden-Württemberg und
nationales Forschungszentrum in der Helmholtz-Gemeinschaft

KIT Scientific Publishing 2013
Print on Demand

ISSN 1865-1100
ISBN 978-3-7315-0056-8

High-Speed, Low-Power and Mid-IR Silicon Photonics Applications

Zur Erlangung des akademischen Grades eines

DOKTOR-INGENIEURS

der Fakultät für Elektrotechnik und Informationstechnik
des Karlsruher Instituts für Technologie

genehmigte

DISSERTATION

von

Dipl.-Phys. Luca Alloatti

aus

Turin, Italien

Tag der mündlichen Prüfung:	20.12.2012
Hauptreferent:	Prof. Dr. sc. nat. Juerg Leuthold
Korreferenten:	Prof. Dr.-Ing. Dr. h. c. Wolfgang Freude
	Prof. Dr.-Ing. Gert F. Trommer

Table of Contents

Abstract (German)

Das Gebiet der integrierten Photonik ist in den letzten Jahren in einer Weise gewachsen, die vergleichbar ist mit der historischen Entwicklung der Elektronik. Sechzig Jahre nach der Erfindung des Transistors gehören Prozessoren mit eine Milliarde Transistoren zu unserem täglichen Leben. Seit den ersten integriert-optischen Strukturen der neunziger Jahre wurden vielfältige neue optische Bauteile wie Modulatoren, Detektoren, Filter, Modenkonverter, optische Schalter, Resonatoren, Verstärker, Laser und noch viele weitere Elemente erfolgreich auf einen gemeinsamen Wafer integriert.

In der Geschichte der Elektronik spielt das Material Silizium eine große Rolle. Silizium kombiniert besondere Eigenschaften, die Massenfabrikation und deswegen preisgünstige Herstellung ermöglichen: Silizium ist eines der häufigsten chemischen Elemente auf der Erde, es kann in großen Einkristallen hergestellt werden, ist ungiftig und mechanisch stabil, und es lässt sich leicht dotieren, ätzen und passivieren. Sein Oxid ist einer der besten elektrischen Isolatoren. Diese positiven Eigenschaften überwiegen Nachteile wie nicht optimale Ladungsträgerbeweglichkeit bei weitem.

Neben den vorteilhaften Eigenschaften im elektronischen Bereich weist Silizium auch außergewöhnliche Vorteile für die Photonik auf. Silizium ist nicht nur transparent bei Telekomunikationwellenlängen und im mittleren Infrarot-Bereich, sondern hat mit die größte Brechzahl, die es bei diesen Wellenlängen gibt. Dies ermöglicht die Konzentration von optischen Feldern auf kleine Raumbereiche und gestattet dadurch die Herstellung hoch-integrierte Wellenleiterstrukturen. Zusätzlich lässt sich die Leitfähigkeit von Silizium durch Ionenimplantation verändern, was ideal für elektrooptische Anwendungen ist. Ferner gestattet die Wahl des Materials Silizium die gleichzeitige Integration von elektrischen und photonischen Bauteilen [1], was eine Vielfalt von Anwendungen ermöglicht.

Silizium wird heutzutage als einziges Materialsystem betrachtet, das Aussichten hat, sich erfolgreich am Markt für integrierte photonische Bauteile zu etablieren [2]. Die Silizium-Photonik ist neben ihrer kommerziellen Bedeutung ein hochattraktives Forschungsgebiet, in dem vieles zu entdecken und zu erfinden bleibt. Ziel dieser Dissertation ist es, neue Bauelemente der Silizium-Photonik zu entwickeln und die grundlegenden physikalischen Effekte zu untersuchen.

Preface

The field of integrated photonics is expanding in the last years in a way which can be compared to the history of electronics, where - sixty years after the invention of the transistor - CPUs with billions of transistors are parts of our daily life. Since the first integrated optical chip at the beginning of the 90s, an optical toolbox comprising modulators, detectors, filters, mode-converters, couplers, all-optical switches, resonators, amplifiers, lasers, and many others, have been demonstrated until today.

In the electronic industry history, silicon plays a major role. Silicon indeed combines several favorable properties, which make it suitable for mass-production and cheapest fabrication. Silicon is abundant, can be grown in large crystals, is mechanically stable, can be easily doped, etched, passivated, its oxide is one of the best existing insulators and it is non-toxic: a rare combination of properties, which are little impaired by its not exceptionally-high carrier mobility.

In addition to the aforementioned characteristics, silicon is an exceptionally favorable material for photonics as well: not only is it transparent in a wide wavelength range which comprises the telecommunication window and the mid-IR range, but most importantly it has one of the highest refractive indices available, enabling the fabrication of sub-micrometer waveguides and small bend radii, which are necessary conditions for high-density integration. Additionally, its electrical characteristics can be tuned by doping or by carrier-injection, making silicon a perfect candidate for electro-optic applications. Moreover, as already shown by the first silicon-photonics company [1], electric and photonic circuits can be successfully integrated on the same silicon chip, multiplying the number of possible applications.

Silicon is considered today the unique material platform which can lead photonics to a successful market [2]. Additionally, silicon photonics is an exciting field of research, where growing governmental and industrial investments are made and where much remains to be discovered and invented.

The aim of this thesis is the development of new "building blocks" for silicon photonics and the study of the physical properties related to them. The thesis is structured as follows:

In Chapter 1, we summarize the theoretical concepts used all-over the thesis. This chapter contains a collection of previously published results, basic equations and models.

In Chapter 2 we present the fastest modulator existing today in silicon photonics. The record-high bandwidth has been achieved by discovering that accumulation layers (induced by a gate voltage) are not only highly conductive, but also have small optical loss. The modulator is based on the combination of a silicon slot-waveguide and an organic nonlinear cladding. The organic cladding requires a special poling procedure which was optimized during this work, and led to driving voltages lower than any other high-speed silicon modulator.

In Chapter 3 we show that second-order nonlinear phenomena can efficiently be exploited in silicon photonics by combining a highly-nonlinear cladding with a dispersion-engineered waveguide. The device is studied from a theoretical point of view. The predicted nonlinear conversion efficiency is by far higher than that of any other third-order nonlinear waveguide demonstrated so far. This device paves the way to the generation of mid-IR wavelengths using low-power CW pumps, and allows for lowest-noise on-chip optical parametric amplification.

In Chapter 4 we demonstrate a phase shifter reacting on a millisecond time-scale, having the smallest voltage-length product $V_\pi L$ and the smallest power consumption ever reported in the silicon photonics community. This result is obtained by combining a silicon slot waveguide with a liquid crystal cladding. This type of device can be exploited in a variety of applications ranging from optical switches to the compensation of fabrication errors where small power consumptions are key features.

In Chapter 5 we determine new physical constants which describe the optical absorption in charge accumulation/inversion layers in silicon. We demonstrate that thin silicon films can be made highly conductive without increasing the optical loss too strongly. The charge layer is induced by a gate voltage. This method is compared to the ion-implant technique, and it is shown that, for practical sheet resistances, smaller optical losses can be obtained.

Chapter 6 finally summarizes the work and gives an outlook for future research. The thesis concludes with appendices, list of publications and the author's curriculum vitae.

Achievements

The following results have been achieved while preparing this thesis:

- First high-speed silicon-organic hybrid (SOH) modulator and first data-transmission experiment at a rate of 42.7 Gbit/s [3]. Previously published results were limited to a bandwidth of 3 GHz [4].
- First silicon-based modulator having a 3 dB bandwidth larger than 100 GHz [5]. This device has a voltage-length product of $V_{\pi}L = 11$ Vmm.
- First proposal of a silicon-organic hybrid waveguide suited for second-order non-linear processes [6].
- First demonstration of a silicon slot-waveguide with liquid crystal cladding [7, 8]. This device established a new record in power consumption and operation voltage for phase shifters in silicon.
- Discovery of small optical losses in accumulation/inversion layers in silicon. We measured for the first time the physical constants describing these losses. We find that the sheet conductance per optical loss figure of merit can be higher than for the ion implant technique [9].
- Measurement of the resistance of accumulation/inversion layers in silicon down to the temperature of 40 mK and gate fields up to 0.25 V/nm [10].
- Modeling, design and measurement of high-speed silicon n channel MOSFETs.
- Investigations on optical nonlinearities in thin films made by atomic-layer deposition (ALD). First experiments indicate the existence nonlinearities of the second order. To the best of our knowledge, this has never been demonstrated in the literature before.
- Construction of a vacuum chamber for on-chip poling of the organic nonlinear cladding.
- Optimization of poling parameters for maximum nonlinearity. This led to the best Pockels coefficient measured at frequencies larger than 10 GHz [5].
- Development of a particle-free anisotropic wet etching process for smooth silicon waveguides [11, 12].
- Numerical investigation of waveguide bends for anisotropic wet etching fabrication [12].
- Design of novel slotted multimode interference couplers (MMI) [13].
- Development of a reliable process for metal deposition. This process was used for creating the electrodes of the 100 GHz modulator.

- Development of a process for in-house fabrication of silicon rib waveguides. The measurement of the optical absorption in accumulation/inversion layers was made on such waveguides.
- In-house development of a silicidation (NiSi) process.
- Investigation of third-order nonlinearities in hydrogenated amorphous silicon [13].
- Design of a low-loss strip-to-slot waveguide converter based on a logarithmic shape. The converter was first fabricated in 2009 [14]. An optimized version was published in [15].
- Three different photonic chip designs with up to 9 layers. One electronic chip design (FETs).

1 Theoretical Background

1.1 Silicon-Organic Hybrid (SOH) Electro-Optic Modulator

In this section we introduce the most important characteristics of electro-optic modulators for future communication systems. We additionally review the methods and guidelines for designing a high-speed, low-voltage, silicon-organic hybrid (SOH) Pockels effect modulator.

1.1.1 Motivation

Electro-optic modulators are used in a large variety of applications. Most of the devices that we interact with are based on some sort of electrical signals. These devices are for example displays, keyboards, touchscreens, microphones and loudspeakers. Also the devices used for storing or transforming information exploit electrical signals: hard-drives, volatile memories, computers are only a few examples. Transporting information, however, is simpler and more effective if optical signals are used instead of electrical signals, as first realized in the late 60s [16]. This is true not only on large scales such from continent to continent, but also on meter or millimeter scales like from rack to rack, from CPU to RAM, or from core to core in future CPUs [2]. If high-quality conversion of an electrical signal into an optical signal is needed, electro-optic modulators have to be used.

In telecommunications, the traffic volume is predicted to increase at a rate of about 30 % per year, mainly due to cloud-computing and high-definition video applications [17]. In order to meet the increasing bandwidth of future network systems, it would be desirable to have modulation capabilities much larger than 10 Gbit/s [18]. Additionally, the internet traffic power consumption in countries like Germany equals 3 % of the total power consumption. This is not yet a large number today, but in ten years, when the traffic will be ten times larger, a power consumption of 30 % of the country's budget will be unacceptable. This requires that the energy consumption per transmitted bit must be reduced significantly. Roadmaps indicate that if an electro-optic modulator today consumes 25 fJ/bit, this number should become 5 fJ/bit in 2020 [2]. Finally, also the spectral efficiency of the optical network is an essential parameter, since higher efficiencies mean that larger amount of data can be transferred without deploying new optical fibers. In order to increase the spectral efficiency, complex data-formats must be used, requiring that both phase and amplitude of an optical carrier can be varied independently. While changing the optical phase with an intensity-modulator is difficult, varying the optical intensity by means of phase-shifters is simple, as it only requires a Mach-Zehnder interferometer (MZI) configuration.

Finally, modulators of future communication systems should be inexpensive and therefore simple to fabricate. Commercial modulators today having 40 Gbit/s capabilities cost a few thousand Euro. Silicon photonics, on the other hand, is a material platform which allows fabricating waveguides at a cost of a few Euro.

As a conclusion, research on modulators should be focused on silicon devices for cost reasons, on phase-shifters for achieving complex data formats, on increasing their bandwidth for achieving higher data-rates, and on decreasing their RF driving voltage for smaller power-consumption. The silicon-organic hybrid (SOH) modulator developed in this thesis is an ideal candidate for meeting all of these requirements.

1.1.2 Pockels Effect

The Pockels effect, also called linear electro-optic effect or Pockels electro-optic effect, produces a variation of the refractive index of an optical medium when an external electric field is applied. Exploiting this phenomenon is the simplest method for building a phase-shifter. The Pockels effect is caused by the second-order optical nonlinearity and reacts therefore on time scales comparable to the period of oscillation of the optical signal itself (fs) [19]; this allows in principle to build ultra-fast optical modulators. A crystalline material having non-vanishing second-order nonlinearity must be non-centrosymmetric, meaning that it is not possible to find a point P (center) such that every atomic position is invariant under the reflection having center P. Using this criterion it is simple to conclude that several crystals, such as silicon, have no second-order nonlinearity. On the other hand however, every polymer is non-centrosymmetric, making the previous criterion useless. A necessary and more useful condition for non-vanishing second-order nonlinearity is that the material is anisotropic.

A linear material, even if anisotropic, can be described by a *permittivity tensor* ε_{ij} relating the electric field E_j and the displacement vector D_i:

$$D_i = \epsilon_0 \epsilon_{ij} E_j \tag{1.1}$$

where we used Einstein's summation convention and ε_0 is the free space permittivity.

It can be shown [20] that for a lossless material ε_{ij} is symmetric and real. As a consequence, it is possible to find an orthogonal (x, y, z) reference system such that ε_{ij} is diagonal. In the following we will assume the indices $i = 1, 2, 3$ or $i = x, y, z$ to be equivalent. Vectors will be indicated with bold fonts.

A second-order nonlinear material, whose Pockels effect is exploited, can be imagined as a linear material whose permittivity tensor can be varied by means of an external electric field. In this thesis we will consider uniquely second-order nonlinear materials consisting of polymer-dispersed chromophores. Prior to changing the molecular orientation of the nonlinear chromophores by a poling procedure, the material is amorphous and isotropic, implying that the dielectric tensor is a multiple of the identity and the second-order nonlinearity is equal to zero. After poling the material by applying an electric field along a certain axis (z) while the material is heated close to its glass transition temperature, the material is invariant under rotations along the poling axis. This means that the dielectric tensor is diagonal with $\varepsilon_{11} = \varepsilon_{22}$.

We will now review how the electro-optic effect is described mathematically, and the constraints which the symmetries of our material imply. We first define the *impermeability tensor* η_{ij}, as the inverse matrix of the permittivity tensor:

$$\eta_{il}\epsilon_{lj} \equiv \delta_{ij} \tag{1.2}$$

where δ_{ij} is Kronecker's delta. It follows:

$$E_i = \frac{1}{\epsilon_0}\eta_{ij}D_j. \tag{1.3}$$

When an external electric field $\mathbf{E}^{(\mathrm{mod})}$ is applied, the impermeability tensor is changed into

$$\eta_{ij} \to \eta_{ij} + \delta\eta_{ij} \tag{1.4}$$

where $\delta\eta_{ij}$ must be real and symmetric as well. The electro-optic tensor is defined by

$$\delta\eta_{ij} \equiv r_{ijk}E_k^{(\mathrm{mod})} \tag{1.5}$$

where we have neglected any higher order term in the electric field. Because of the symmetry of $\delta\eta_{ij}$, we find

$$r_{ijk} = r_{jik} \quad \forall k. \tag{1.6}$$

If no further symmetries are exploited, the electro-optic tensor has $6\times3 = 18$ independent components. Because the symmetry of the permittivity tensor reduces the number of independent components of the electro-optic tensor from 27 to 18, it is common practice to define a 6×3 matrix r_{ij}, whose k^{th} column consists of the elements r_{ijk} chosen in a conventional order [19]:

$$r_{ij} = \begin{pmatrix} r_{111} & r_{112} & r_{113} \\ r_{221} & r_{222} & r_{223} \\ r_{331} & r_{332} & r_{333} \\ r_{231} & r_{232} & r_{233} \\ r_{131} & r_{132} & r_{133} \\ r_{121} & r_{122} & r_{123} \end{pmatrix}. \tag{1.7}$$

We will now show that if the polymer is poled along the 3 direction, then $r_{ijk} = 0$ for $k = 1, 2$. The nonlinear polymer is indeed invariant under rotations along the poling axis, meaning in particular that by applying a modulating field in the plane perpendicular to the poling axis, or a field having same modulus but opposite direction, $\delta\eta_{ij}$ must be same

$$r_{ijk}\,E_k^{(\mathrm{mod})} = r_{ijk}\,(-)E_k^{(\mathrm{mod})} \quad \forall i,j \quad k = 1,2 \tag{1.8}$$

which proves the proposition.

Let's now consider the consequences of polymer's rotational symmetry. To this end we consider a rotation of π/2 around the 3-axis which is described by

$$R = \begin{pmatrix} 0 & 1 & 0 \\ -1 & 0 & 0 \\ 0 & 0 & 1 \end{pmatrix}$$

Invariance of the material under such a rotation means that if an optical field **E** traverses the material producing the displacement $\mathbf{D} = \epsilon_0 \epsilon \mathbf{E}$, then the displacement **D'** produced by the rotated field $R\mathbf{E}$ must be equal to the original displacement **D** rotated by R, i.e. $R\mathbf{D}$. Mathematically

$$\mathbf{D} = \epsilon_0 \epsilon \mathbf{E} \quad \Leftrightarrow \quad R\mathbf{D} = \epsilon_0 \epsilon R\mathbf{E} \quad \forall \mathbf{E} \tag{1.9}$$

The latter equation implies

$$\epsilon R = R\epsilon, \tag{1.10}$$

which is equivalent to

$$R\eta = \eta R. \tag{1.11}$$

From Eqs. (1.4) and (1.11) it follows

$$R_{im}\,\delta\eta_{mj} = \delta\eta_{im}\,R_{mj}. \tag{1.12}$$

If we now turn on a modulating field along the 3-axis, we get

$$R_{im}\,r_{mj3} = r_{im3}\,R_{mj}. \tag{1.13}$$

The latter equation implies that r_{ij3} takes the following form

$$r_{ij3} = \begin{pmatrix} r_{113} & 0 & 0 \\ 0 & r_{113} & 0 \\ 0 & 0 & r_{333} \end{pmatrix} \tag{1.14}$$

Finally we obtain the r_{ij} matrix for a second-order nonlinear polymer poled along the 3-axis

$$r_{ij} = \begin{pmatrix} 0 & 0 & r_{113} \\ 0 & 0 & r_{113} \\ 0 & 0 & r_{333} \\ 0 & 0 & 0 \\ 0 & 0 & 0 \\ 0 & 0 & 0 \end{pmatrix}. \tag{1.15}$$

Equation (1.15) shows that a poled nonlinear polymer can be described by only two independent numbers, namely $r_{13} \equiv r_{113}$ and $r_{33} \equiv r_{333}$.

Summarizing the results obtained so far: If a nonlinear polymer is poled along the 3-axis, an external modulating field $(E_1^{(\text{mod})}, E_2^{(\text{mod})}, E_3^{(\text{mod})})$ changes the impermeability tensor in the following way:

$$\begin{pmatrix} \eta_{11} & 0 & 0 \\ 0 & \eta_{11} & 0 \\ 0 & 0 & \eta_{33} \end{pmatrix} \rightarrow \begin{pmatrix} \eta_{11} & 0 & 0 \\ 0 & \eta_{11} & 0 \\ 0 & 0 & \eta_{33} \end{pmatrix} + \begin{pmatrix} r_{113} & 0 & 0 \\ 0 & r_{113} & 0 \\ 0 & 0 & r_{333} \end{pmatrix} E_3^{(\text{mod})} \tag{1.16}$$

Here the modulating fields along the 1-axis and 2-axis do not play a role.

If a plane wave propagating in the nonlinear polymer has the electric field oriented along the k-axis ($k = 1,2,3$), it experiences an index of refraction

$$n_k + \delta n_k = \sqrt{\frac{1}{\eta_{kk} + r_{kk3}E_3^{(\mathrm{mod})}}} \approx n_k - \frac{1}{2} n_k^3 r_{kk3} E_3^{(\mathrm{mod})} \tag{1.17}$$

where $n_k \equiv \sqrt{\epsilon_{kk}} = \sqrt{1/\eta_{kk}}$ and higher orders of the modulating field have been neglected.

Finally, we observe that it is usually assumed for nonlinear polymer-dispersed chromophores that the r_{33} coefficient is dominant over all the others [21, 22]. To the best of our knowledge however, this still lacks experimental verification. Still, for the sake of simplicity, we will assume from now on that the r_{33} is the only electro-optic coefficient different from zero.

1.1.3 Bandwidth of a Modulator

In the silicon photonics literature, the bandwidth of an electro-optic modulator is not always defined in the same way [4, 18]. We observe that if a Mach-Zehnder (MZ) modulator is connected to a CW laser source and it is biased such that the output optical power is ½ of its maximum value (quadrature point), the variation of optical power ΔP_{out} is linear in the variation of the voltage ΔV_{in} applied to it (in the small signal limit). If the intensity modulation is detected with a photodetector, the RF power generated by the photodetector is quadratic in the AC voltage applied to the modulator. We define the electro-optic responsivity of a MZ modulator as [23]

$$R(\mathrm{dB}) \equiv 20\log_{10}\left(\frac{\Delta P_{\mathrm{out}}}{1\,\mathrm{W}}\frac{1\,\mathrm{V}}{\Delta V_{\mathrm{out}}}\right) \equiv 20\log_{10}\left(\frac{\Delta P_{\mathrm{out}}}{1\,\mathrm{W}}\frac{1\,\mathrm{V}}{\Delta V_{\mathrm{out}}}\right). \tag{1.18}$$

The electro-optic EO bandwidth of a MZ modulator is the frequency at which its responsivity expressed in dB is reduced by a certain constant. Commonly used constants are 3 dB [24, 25] or 6 dB [4]. MZ modulators sold as 40 Gbit/s devices have a specified 3 dB bandwidth of 25 GHz [25].

Finally, for a phase modulator, it is possible to define its bandwidth imagining that it is mounted inside one arm of a MZ interferometer. The frequency at which its modulation index (which is the maximum phase-shift achieved) is reduced by a factor $\sqrt{2}$ corresponds to the 3 dB EO bandwidth, while the point at which it is reduced by a factor 2 corresponds to the 6 dB EO bandwidth.

The bandwidth of our modulators is mainly limited by their RC time constant, as discussed in Section 2.2. A second important bandwidth limitation is the frequency-dependent propagation loss of the electrodes. These problems are studied and described in [26].

1.1.4 Driving Voltage

The driving voltage is one of the most important characteristics of an electro-optic modulator. In this thesis we will consider uniquely the geometry shown in Fig. 1. The waveguide consists of a slot waveguide connected to metal electrodes by means of thin silicon strips (strip-loaded slot waveguide). The waveguide is coated with a nonlinear material which is poled by applying a voltage between the two silicon strips. The optical mode used is the TE_{00}, which has the main field component parallel to the poling direction.

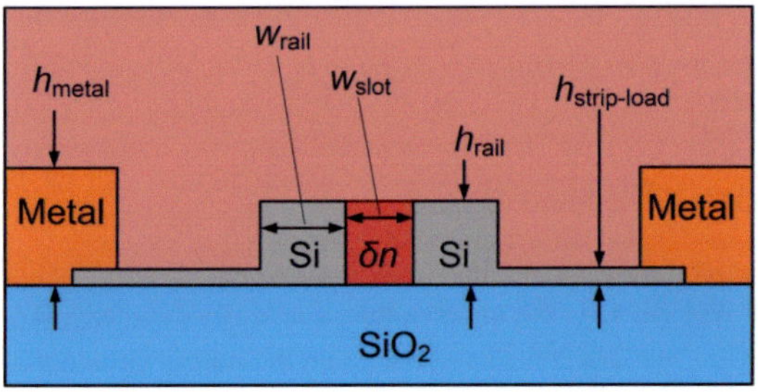

Fig. 1. Cross section of a strip-loaded slot waveguide modulator.

The optical mode propagates in the waveguide with a propagation constant $\beta \equiv n_{\mathrm{eff}} 2\pi / \lambda_0$, where n_{eff} is the effective index of the mode and λ_0 is the free-space wavelength. If a voltage $V^{(\mathrm{mod})}$ is applied across the slot, the quasi-TE_{00} mode experiences a variation in the refractive index of the polymer, see Eq. (1.17),

$$\delta n = -\frac{1}{2} n^3 r_{33} V^{(\mathrm{mod})} / w_{\mathrm{slot}}. \tag{1.19}$$

As a consequence, the effective index of the waveguide is changed, $n_{\mathrm{eff}} \to n_{\mathrm{eff}} + \delta n_{\mathrm{eff}}$. The ratio between the change of the effective index and the change of the polymer refractive index defines the interaction factor Γ,

$$\delta n_{\mathrm{eff}} \equiv \Gamma \delta n. \tag{1.20}$$

The interaction factor can be calculated numerically by means of an overlap integral [27]. A faster method derives Γ by simulating the modes [28] for two slightly different refractive indices in the slot, and calculating the difference of the corresponding effective indices. If V_π denotes the voltage required for obtaining a phase shift of π, and L is the device length, the voltage-length product $V_\pi L$ is a quantity which very well describes the performance of a modulator:

$$V_\pi L = \frac{\lambda_0 w_{\mathrm{slot}}}{n^3 r_{33} \Gamma}. \tag{1.21}$$

It is interesting to observe that in equation (1.21) appears the ratio $\Gamma / w_{\mathrm{slot}}$. It can be shown [26] that this value increases monotonically and converges to a finite value if the slot width w_{slot} tends to zero [26]. In general, therefore, smaller slots are preferable. However, with a standard 193 nm DUV lithography system, the slot size cannot be made smaller than 100 nm leading to a Γ / w_{slot} roughly a factor 10 smaller than the optimum.

For a slot size $w_{\text{slot}} = 120\,\text{nm}$, a strip height $h_{\text{rail}} = 220\,\text{nm}$ (as standard in silicon photonics) a strip-load height $h_{\text{strip-load}} = 70\,\text{nm}$ (as fixed by the silicon photonic fabs IMEC and LETI), a cladding refractive index $n = 1.682$ and a wavelength $\lambda_0 = 1550\,\text{nm}$ the interaction factor Γ in Eq. (1.21) is maximized for a strip width $w_{\text{rail}} = 240\,\text{nm}$ [29] resulting in $\Gamma = 0.196$. Assuming finally an electro-optic coefficient $r_{33} = 70\,\text{pm/V}$, equation (1.21) leads to $V_\pi L = 2.85\,\text{Vmm}$.

1.2 Interaction Factors in a High-Index-Contrast Waveguide

In this Section we derive equations which allow calculating the absorption, the amplification or the generation of optical signals inside a waveguide. The formalism used here is valid also for high-index-contrast waveguides, such as silicon waveguides surrounded by glass or polymers. The theory presented here will be used in the following Chapters when analyzing free-carrier absorption in accumulation/inversion layers, and second-order nonlinear waveguides.

1.2.1 Excitation of Modes by Localized Currents

In this section we report a basic equation which allows describing the evolution of mode amplitudes in a waveguide under the influence of electromagnetic sources. The content of this section follows the arguments of the book of Vassallo [30].

In this section we will adopt the convention of indication with a tilde (e.g. $\tilde{\mathbf{E}}$) the Fourier coefficients of the terms $e^{-i\omega t}$., where ω is *positive*. Coefficients of the terms $e^{i\omega t}$ (ω is *positive*) will consequently be indicated with the complex conjugate $\tilde{\mathbf{E}}^*$. For the electric field $\mathbf{E}$ we will write for example (cc. represents the complex conjugate of the foregoing expression):

$$\mathbf{E}(t) = \sum_{\omega} \left(\tilde{\mathbf{E}}(\omega) e^{-i\omega t} + \text{cc.} \right). \tag{1.22}$$

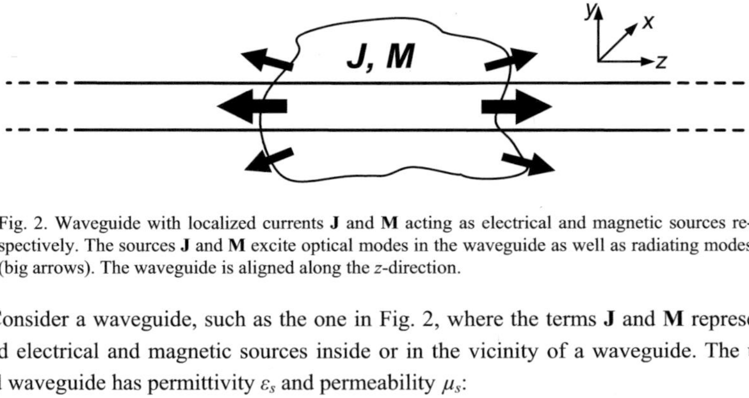

Fig. 2. Waveguide with localized currents **J** and **M** acting as electrical and magnetic sources respectively. The sources **J** and **M** excite optical modes in the waveguide as well as radiating modes (big arrows). The waveguide is aligned along the *z*-direction.

Consider a waveguide, such as the one in Fig. 2, where the terms **J** and **M** represent localized electrical and magnetic sources inside or in the vicinity of a waveguide. The unperturbed waveguide has permittivity ε_s and permeability μ_s:

$$\begin{cases} \operatorname{rot} \mathbf{E} = -\mu_s \dfrac{\partial \mathbf{H}}{\partial t} - \mathbf{M} \\ \operatorname{rot} \mathbf{H} = \epsilon_s \dfrac{\partial \mathbf{E}}{\partial t} + \mathbf{J} \end{cases} \tag{1.23}$$

Because of the linearity of Maxwell equations, for every given angular frequency ω it must hold:

$$\begin{cases} \text{rot}\,\widetilde{\mathbf{E}} = i\omega\mu_s \widetilde{\mathbf{H}} - \widetilde{\mathbf{M}} \\ \text{rot}\,\widetilde{\mathbf{H}} = -i\omega\epsilon_s \widetilde{\mathbf{E}} + \widetilde{\mathbf{J}} \end{cases} \tag{1.24}$$

In the remaining of this section we will therefore consider only monochromatic signals. For definiteness, let us further assume that the waveguide is oriented along the z-direction. If the sources **J** and **M** are set to zero, the guided modes propagating in the positive direction are described by:

$$\mathbf{E}^{(\nu)}(x,y,z,t) = A^{(\nu)} \begin{pmatrix} \mathcal{E}_x^{(\nu)}(x,y) \\ \mathcal{E}_y^{(\nu)}(x,y) \\ \mathcal{E}_z^{(\nu)}(x,y) \end{pmatrix} e^{-i\omega t + i\beta^{(\nu)} z} + \text{cc.} \tag{1.25}$$

$$\mathbf{H}^{(\nu)}(x,y,z,t) = A^{(\nu)} \begin{pmatrix} \mathcal{H}_x^{(\nu)}(x,y) \\ \mathcal{H}_y^{(\nu)}(x,y) \\ \mathcal{H}_z^{(\nu)}(x,y) \end{pmatrix} e^{-i\omega t + i\beta^{(\nu)} z} + \text{cc.} \tag{1.26}$$

where $\nu > 0$ is the mode index, A is a dimensionless amplitude, ω is the angular frequency and β is the propagation constant.

We will indicate modes propagating in the negative direction with $-\nu$. It can be shown [30] that these are related to the modes propagating in the positive direction by:

$$\mathbf{E}^{(-\nu)}(x,y,z,t) = A^{(-\nu)} \begin{pmatrix} \mathcal{E}_x^{(-\nu)}(x,y) \\ \mathcal{E}_y^{(-\nu)}(x,y) \\ \mathcal{E}_z^{(-\nu)}(x,y) \end{pmatrix} e^{-i\omega t + i\beta^{(-\nu)} z} + \text{cc.} = A^{(-\nu)} \begin{pmatrix} \mathcal{E}_x^{(\nu)}(x,y) \\ \mathcal{E}_y^{(\nu)}(x,y) \\ -\mathcal{E}_z^{(\nu)}(x,y) \end{pmatrix} e^{-i\omega t - i\beta^{(\nu)} z} + \text{cc.} \tag{1.27}$$

$$\mathbf{H}^{(-\nu)}(x,y,z,t) = A^{(-\nu)} \begin{pmatrix} \mathcal{H}_x^{(-\nu)}(x,y) \\ \mathcal{H}_y^{(-\nu)}(x,y) \\ \mathcal{H}_z^{(-\nu)}(x,y) \end{pmatrix} e^{-i\omega t + i\beta^{(-\nu)} z} + \text{cc.} = A^{(-\nu)} \begin{pmatrix} -\mathcal{H}_x^{(\nu)}(x,y) \\ -\mathcal{H}_y^{(\nu)}(x,y) \\ \mathcal{H}_z^{(\nu)}(x,y) \end{pmatrix} e^{-i\omega t - i\beta^{(\nu)} z} + \text{cc.} \tag{1.28}$$

If we now turn on the sources **J** and **M**, we can describe the electromagnetic field in the waveguide by replacing the constant amplitudes $A^{(\pm\nu)}$ in the equations above with space-dependent amplitudes $A^{(\pm\nu)}(z)$.

It can be shown that the variation of the $A^{(\pm\nu)}(z)$ amplitudes along the z-direction is given by [30]:

$$\frac{\partial A^{(\nu)}(z)}{\partial z} = -\frac{1}{2\int \mathrm{d}x\mathrm{d}y \left(\mathcal{E}_x^{(\nu)} \mathcal{H}_y^{(\nu)} - \mathcal{E}_y^{(\nu)} \mathcal{H}_x^{(\nu)} \right)} \int \mathrm{d}x\mathrm{d}y \left(\widetilde{\mathbf{J}} \cdot \mathcal{E}^{(-\nu)} - \widetilde{\mathbf{M}} \cdot \mathcal{H}^{(-\nu)} \right) e^{-i\beta^{(\nu)} z}. \tag{1.29}$$

1.2.2 Generation of a Guided Optical Wave by Second-Order Nonlinear Mixing

In this Section we consider the case of second-order nonlinear mixing in a high-index contrast waveguide. The second-order nonlinear tensor can have arbitrary form and be non-constant over the waveguide cross-section. We will determine the equations which govern the evolution of the light waves involved by applying Eq. (1.29) to the case where the source **J** is generated by a second-order nonlinear term (the magnetic source **M** is equal to zero in this Section).

For simplicity, we will first ignore that the nonlinear susceptibility tensors might have frequency dependence. We will further assume that if an electrical signal **E**(t) impinges on a nonlinear medium, the material reacts by creating a polarization **P**(t) which can be expanded in a power series of the electrical signal [19]:

$$P_i(t) = \epsilon_0 \left[\chi_{ij}^{(1)} E_j(t) + \chi_{ijk}^{(2)} E_j(t) E_k(t) + \chi_{ijkl}^{(3)} E_j(t) E_k(t) E_l(t) + \ldots \right] \tag{1.30}$$

where the $\chi^{(1)}$ is the (linear) susceptibility of the material (which determines the material's index of refraction), $\chi^{(2)}$ is the second-order nonlinear susceptibility tensor, $\chi^{(3)}$ is the third-order nonlinear susceptibility tensor and both **P** and **E** are real physical fields. We further assumed Einstein's summation convention (no restriction $i < j$ or alike is made in the summation). In Eq. (1.30), no time convolution appears, because of the assumption that the susceptibility tensors do not depend on frequency. Strictly speaking, this assumption is valid only for vacuum, but it makes the coming derivations simpler [31] and will be abandoned later on.
In the following we will consider the case of difference-frequency generation (DFG), where pump (angular frequency ω_2, mode ν_2) and signal (angular frequency ω_1, mode ν_1) generate the idler wavelength (angular frequency $\omega_3 = \omega_2 - \omega_1$, mode ν_3). To this end, we consider the following superposition of monochromatic waves:

$$\mathbf{E}(t) = \widetilde{\mathbf{E}}(\omega_1) e^{-i\omega_1 t} + \widetilde{\mathbf{E}}(\omega_2) e^{-i\omega_2 t} + \text{cc.} \tag{1.31}$$

and insert it into Eq. (1.30). If we collect only the terms having angular frequency $\omega_2 - \omega_1$ or $\omega_1 - \omega_2$, i.e. the difference-frequency generation (DFG) terms, we find:

$$P_i(t)\,|_{\text{DFG}} = 2\epsilon_0 \chi_{ijk}^{(2)} \left[\tilde{E}_j^*(\omega_1) \tilde{E}_k(\omega_2) e^{-i(\omega_2 - \omega_1)t} + \text{cc.} \right]. \tag{1.32}$$

If we now separate the polarization in Eq. (1.30) in a linear and a nonlinear (NL) term:

$$P_i(t) = \epsilon_0 \chi_{ij}^{(1)} E_j(t) + P_i^{(\text{NL})}(t) \tag{1.33}$$

the fourth Maxwell equation can be rewritten as [19]:

$$\text{rot}\,\mathbf{H}\,|_i = \epsilon_0 \chi_{ij}^{(1)} \frac{\partial E_j}{\partial t} + \frac{\partial P_i^{(\text{NL})}}{\partial t}. \tag{1.34}$$

Comparison with Eq. (1.23) shows that the source **J** is given by the time derivative on the nonlinear polarization:

$$\mathbf{J} = \frac{\partial \mathbf{P}^{(\mathrm{NL})}}{\partial t} = -2i(\omega_2 - \omega_1)\epsilon_0 \chi_{ijk}^{(2)} \tilde{E}_j^*(\omega_1)\tilde{E}_k(\omega_2)e^{-i(\omega_2-\omega_1)t} + \mathrm{cc.}\,,$$

from which, by making use of Eq. (1.32), we find an explicit expression of the source term $\tilde{\mathbf{J}}$ appearing in the equation (1.29) of Vassallo:

$$\tilde{J}_i(\omega_2 - \omega_1) = -2i(\omega_2 - \omega_1)\epsilon_0 \chi_{ijk}^{(2)} \tilde{E}_j^*(\omega_1)\tilde{E}_k(\omega_2). \tag{1.35}$$

We now substitute the fields in the equation above with the mode-field distributions, Eq. (1.25):

$$\tilde{J}_i(\omega_2 - \omega_1) = -2i(\omega_2 - \omega_1)\epsilon_0 \chi_{ijk}^{(2)} A^*(\omega_1)A(\omega_2)\left[\mathcal{E}_j^{(\nu_1)}(\omega_1)\right]^* \mathcal{E}_k^{(\nu_2)}(\omega_2) \tag{1.36}$$

Finally, inserting this expression in the equation (1.35) of Vassallo we find for the idler amplitude:

$$\frac{\partial A^{(\nu_3)}(\omega_3, z)}{\partial z} = i\frac{\epsilon_0 \omega_3 A^*(\omega_1)A(\omega_2)}{\int \mathrm{d}x\mathrm{d}y\left(\mathcal{E}_x^{(\nu_3)}\mathcal{H}_y^{(\nu_3)} - \mathcal{E}_y^{(\nu_3)}\mathcal{H}_x^{(\nu_3)}\right)}\int \mathrm{d}x\mathrm{d}y\, \chi_{ijk}^{(2)}\left[\mathcal{E}_j^{(\nu_1)}(\omega_1)\right]^* \mathcal{E}_k^{(\nu_2)}(\omega_2)\, \mathcal{E}_i^{(-\nu_3)}(\omega_3)e^{i\Delta\beta z}$$
(1.37)

$$\omega_3 \equiv \omega_2 - \omega_1 \tag{1.38}$$

$$\Delta\beta \equiv \left(\beta^{(\nu_2)} - \beta^{(\nu_1)}\right) - \beta^{(\nu_3)} \tag{1.39}$$

Equation (1.37) allows calculating the generation of a signal generated by DFG in a strongly-guiding optical waveguide. The back-propagating mode $-\nu_3$ in Eq. (1.37) can be expressed in term of forward-propagating modes thanks to Eq. (1.27). We finally observe that the space-dependent nonlinear susceptibility in Eq. (1.35) can be considered frequency dependent as well: $\chi_{ijk}^{(2)} = \chi_{ijk}^{(2)}(\mathbf{r}, \omega_2 - \omega_1, \omega_2, \omega_1)$. We further notice that Eq. (1.38) represents the energy conservation of the photons involved, and that Eq. (1.39) describes the momentum mismatch between the different optical waves. If the momentum mismatch is zero, then the well-known phase-matching condition is satisfied [19]. The result in Eq. (1.37) can be easily generalized to any other three-wave mixing process.

1.2.3 Attenuation of a Guided Optical Wave

We now apply Eq. (1.29) to the case that dielectric loss is added to a lossless waveguide. In this Section we assume again that the waveguide is oriented along the z-axis of an (x, y, z) coordinate system and that the waveguide has no magnetic sources, $\mathbf{M} = 0$. Let us consider a waveguide defined by following permittivity profile:

$$\epsilon_s(x, y) = \epsilon_0 n_s^2(x, y) \tag{1.40}$$

where $n_s(x, y)$ is the refractive index of the waveguide. We now introduce loss in the system by adding the following (complex) perturbation to the permittivity:

$$\epsilon_s(x, y) \rightarrow \epsilon_s(x, y) + \delta\epsilon(x, y, z)\,. \tag{1.41}$$

We observe that the Maxwell equations (1.24) describe the new system, if we set:

$$\tilde{\mathbf{J}} \equiv -i\omega\delta\epsilon\,\tilde{\mathbf{E}}\,. \tag{1.42}$$

We now define a local loss coefficient $\alpha(x,y,z)$ by the following equation (c is the speed of light):

$$\delta\epsilon(x,y,z) \equiv i\frac{\alpha(x,y,z)n_s(x,y)c}{\omega}\epsilon_0\,. \tag{1.43}$$

It can be shown that in the case of a plane wave propagating in a homogeneous medium (permittivity constant in space), the coefficient α defined in Eq. (1.43) correctly describes the power attenuation of the plane wave [31].

Combining Eqs. (1.29), (1.42) and (1.43) we finally obtain that the attenuation α_{wg} of a mode v propagating in the waveguide is given by:

$$\alpha_{\text{wg}} = -\frac{2}{A^{(v)}}\frac{\partial A^{(v)}(z)}{\partial z} = \frac{c\epsilon_0}{\int \mathrm{d}x\mathrm{d}y\left(\mathcal{E}_x^{(v)}\mathcal{H}_y^{(v)} - \mathcal{E}_y^{(v)}\mathcal{H}_x^{(v)}\right)}\int \mathrm{d}x\mathrm{d}y\, n_s(x,y)\alpha(x,y,z)\mathcal{E}^{(v)}\cdot\mathcal{E}^{(-v)} \tag{1.44}$$

which can be simplified using Eq. (1.27) into:

$$\alpha_{\text{wg}} = \frac{c\epsilon_0}{\int \mathrm{d}x\mathrm{d}y\left(\mathcal{E}_x^{(v)}\mathcal{H}_y^{(v)} - \mathcal{E}_y^{(v)}\mathcal{H}_x^{(v)}\right)}\int \mathrm{d}x\mathrm{d}y\, n_s(x,y)\alpha(x,y,z)\left[\left(\mathcal{E}_x^{(v)}\right)^2 + \left(\mathcal{E}_y^{(v)}\right)^2 - \left(\mathcal{E}_z^{(v)}\right)^2\right]. \tag{1.45}$$

This equation allows calculating the propagation loss of a mode inside a strongly-guiding waveguide in terms of the local loss of the material. For a plane wave propagating in a homogeneous medium having loss α, Eq. (1.45) implies correctly that $\alpha_{\text{wg}} = \alpha$

1.3 Metal-Insulator-Semiconductor (MIS) Model

In this Section we study the electron- and hole-concentration in the vicinity of a Si/SiO_2 interface under the influence of an external electric DC field (generated for example by a gate voltage). This gives information on the resistance, the thickness, and dependence on the external electric field. These parameters are key for the interpretation of the performance of the 100 GHz silicon electro-optic modulator developed in Section 2.3. Moreover, the analysis here presented provides information concerning the thickness and the maximal charge-density at the interface, which is used in Chapter 5 for studying the optical loss of accumulation/inversion layers.

Most of the results inferred from this section are based on the numerical integration of a first-order nonlinear differential equation, Eq. (1.53). While deriving this equation, some approximations are made. If accurate numerical values were needed, we used the industry-standard numerical solver ATLAS, finding astonishingly small discrepancies with the predictions of the MIS model [32]. The MIS model therefore remains in all cases not only a valid tool for the qualitative interpretations, but provides very accurate predictions also when driven beyond the limits determined by its hypothesis.

1.3.1 Definition of the Model

The model reported here is standard in solid-state physics and is repeated here for the convenience of the reader. The Metal-Insulator-Semiconductor (MIS) structure is shown in Fig. 3, where d is the thickness of the insulator, V_{gate} is the voltage applied across the insulator, E_c is the conduction band edge, E_v is the valence band edge, E_F is the Fermi energy. The voltage V_{gate} is defined to be *positive* if the semiconductor is grounded and the metal is connected to a positive voltage source. In other words, V_{gate} is the potential of the metal if the semiconductor is grounded.

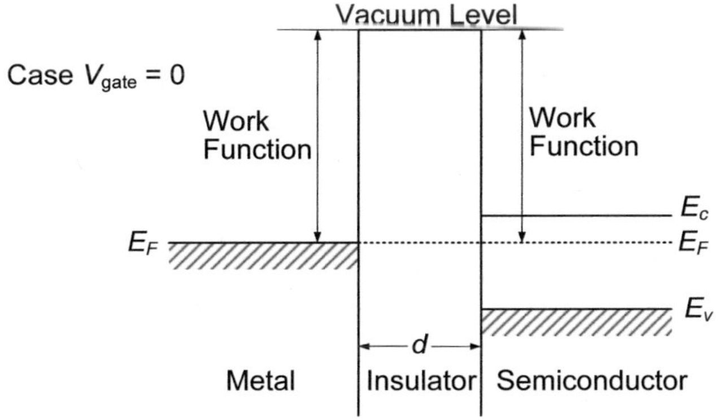

Fig. 3: Metal-Insulator-Semiconductor (MIS) structure. When no gate voltage V_{gate} is applied across the insulator the bands are flat (ideal MIS). Picture reproduced after [33].

We further make the following assumptions:

1. The work-functions (difference between Femi level and vacuum level) of the semiconductor and of the metal are identical (when the two materials are remote from each other and do not necessarily form a single thermodynamic system sharing the same Fermi energy). This means that in the MIS configuration and for $V_{gate} = 0$ the bands are flat (flat-band condition)
2. There are no surface charges on the insulator. Therefore the unique charges which can exist in the system are those in the semiconductor and those (equal but opposite sign) in the metal close to the insulator.
3. The insulator is perfect, i.e. there is no charge transport in the insulator.
4. The semiconductor is homogeneous, meaning that its dielectric constant is constant over space.
5. The semiconductor is *non-degenerate*, meaning that the Fermi energy lies within the energy gap and is separated by several $k_B T$ (say 5× or more) from both the band edges (k_B is Boltzmann's constant, T is the absolute temperature). When this condition is satisfied, the Fermi-Dirac distribution function $f(E)$, as well as the complementary distribution $1\text{-}f(E)$ can be simplified, resulting in the corresponding Maxwell-Boltzmann exponential distributions [33]:

$$f(E) = \frac{1}{e^{(E-E_F)/k_B T} + 1} \approx \frac{1}{e^{(E-E_F)/k_B T}} = e^{-(E-E_F)/k_B T} \tag{1.46}$$

$$1 - f(E) = \frac{1}{e^{(E_F-E)/k_B T} + 1} \approx \frac{1}{e^{(E_F-E)/k_B T}} = e^{-(E_F-E)/k_B T} \tag{1.47}$$

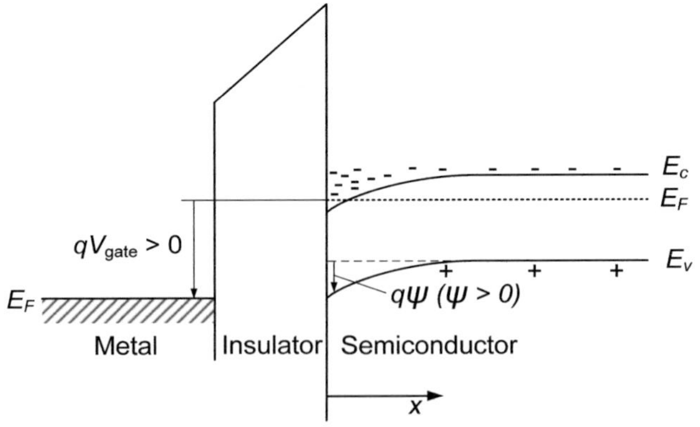

Fig. 4: Energy-band diagram of a MIS structure with positive gate voltage V_{gate} applied and n-type semiconductor. Electrons are accumulated at the insulator-semiconductor interface. Electrons and holes are represented with (-) and (+) signs respectively. The (positive) elementary charge is indicated with q. Picture reproduced after [33].

For simplicity we assume here that the semiconductor is *n*-type, as usually the case for the devices considered in this thesis. When a bias V_{gate} is applied across the insulator, three cases are to be considered. For $V_{\text{gate}} > 0$, the bands in the semiconductor bend downwards, and a negative charge layer forms of at the insulator-semiconductor interface. Since the semiconductor is *n*-doped, we speak of *accumulation*. In the case that $V_{\text{gate}} < 0$, the majority carriers are *depleted.* If a larger (in modulus) negative voltage is applied, the Fermi level might get closer to the valence band than to the conduction band. At this point the number of holes (minority carriers) at the surface is similar or larger than the number of electrons; in this case we speak of *inversion*. The case of $V_{\text{gate}} > 0$ and *n*-type semiconductor is shown in Fig. 4.

The potential energy $q\psi$ (q is the positive elementary charge) is defined to be zero in the bulk of the semiconductor and measures the deflection of the conduction or valence band edges of the semiconductor. It is defined to be *positive* when the bands bend downwards. The electron and hole concentrations as a function of the potential energy are given by the following relations ($N_{e,0}$ and $N_{h,0}$ the carrier concentrations far away from the interface):

$$N_e = N_{e,0} e^{+q\psi/k_B T} \tag{1.48}$$

$$N_h = N_{h,0} e^{-q\psi/k_B T} \tag{1.49}$$

The derivative of the potential ψ gives the electric field inside the semiconductor (notice that there is no minus sign because of the sign definition of ψ):

$$\mathcal{E}_x = \frac{\partial \psi}{\partial x}$$

which, combined with the Maxwell equations of macroscopic electromagnetism, leads to the one-dimensional Poisson equation (ε_s is the dielectric constant of the semiconductor):

$$\frac{\partial^2 \psi}{\partial x^2} = \frac{\rho(x)}{\epsilon_s} \tag{1.50}$$

where $\rho(x)$ is the free space-charge density (with material polarization excluded) given by:

$$\rho(x) = q\left(N_D - N_A + N_h - N_e\right) \tag{1.51}$$

where N_D and N_A are the donor and acceptor densities, respectively. The resulting Poisson equation is therefore:

$$\frac{\partial^2 \psi}{\partial x^2} = \frac{q}{\epsilon_s}\left[N_{h,0}\left(e^{-q\psi/k_B T} - 1\right) - N_{e,0}\left(e^{q\psi/k_B T} - 1\right)\right] \tag{1.52}$$

which leads, after integration by parts, to:

$$\left(\frac{\partial \psi}{\partial x}\right)^2 = \frac{2k_B T N_{e,0}}{\epsilon_s}\left[\left(e^{q\psi/k_B T} - \frac{q\psi}{k_B T} - 1\right) + \frac{N_{h,0}}{N_{e,0}}\left(e^{-q\psi/k_B T} + \frac{q\psi}{k_B T} - 1\right)\right] \tag{1.53}$$

The latter equation can be integrated numerically for different gate voltages and doping levels, leading to the potential $\psi(x)$ which allows calculating, via Eqs. (1.48) and (1.49), the space-dependent electron and hole densities $N_{e,0}(x)$ and $N_{h,0}(x)$.

In the rest of the thesis, we further assume that the insulator thickness is much larger than the space-charge region induced by the gate voltage. This is valid for all the devices considered in this thesis, where the insulator thickness is usually $d = 2$ µm (the space-charge region is always at most a few nanometers thick). Under this assumption, the electric field in the insulator can be conveniently approximated by

$$\mathcal{E}_x\left(x=0^-\right)=\frac{V_{gate}}{d}, \tag{1.54}$$

meaning that in the semiconductor close to the interface there is an electric field given by

$$\mathcal{E}_x\left(x=0^+\right)=\frac{\epsilon_i}{\epsilon_s}\frac{V_{gate}}{d}, \tag{1.55}$$

where ε_i is the dielectric constant of the insulator.

1.3.2 Electron and Hole Densities for Different Temperatures and Gate Fields

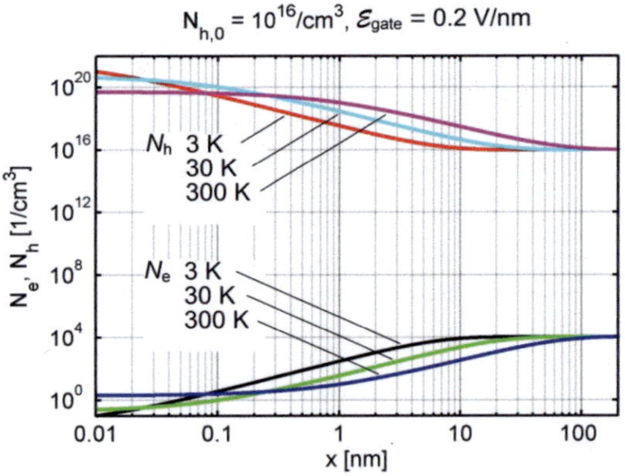

Fig. 5. Accumulation in p-type silicon. Electron and hole density vs. distance for three different temperatures (legend) and gate field $V_{gate}/d = -0.2$ V/nm. Silicon is assumed to have a bulk hole concentration $N_{h,0} = 10^{16}$ cm^{-3}. Since holes are the majority carriers, this case represents hole accumulation. As the temperature is lowered the thickness of the charge layer is decreased.

In this Section Eq. (1.53) is solved numerically for a few special cases. The numerical code used for integration is reported in the Appendix.

We first consider the distribution of electrons and holes for given gate fields and different temperatures. For definiteness, we consider p-type silicon with a bulk hole concentration $N_{h,0} = 10^{16}$ cm^{-3}. The result for negative gate fields are summarized in Fig. 5, while for a positive gate field (inversion) the outcome is shown in Fig. 6. As the temperature is lowered, the thickness of the charge layer decreases. The electron and hole concentrations in Fig. 5 and Fig. 6 vary over several orders of magnitude and on scales up to 100 nm. Still, most of the charges are located in the first few nm from the interface, as it can better be seen by plotting the net charge density on a linear scale This is done in Fig. 7, for different gate fields and at the temperature T =300 °K.

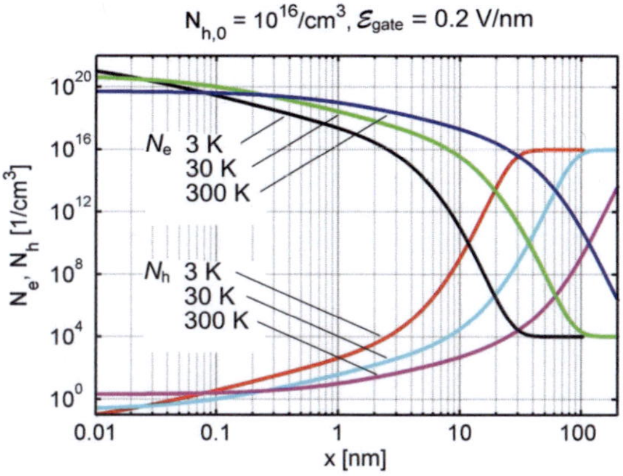

Fig. 6. Inversion in p-type silicon. Electron and hole density vs. distance for three different temperatures (legend) and gate field $V_{gate}/d = 0.2$ V/nm. Silicon is assumed to have a bulk hole concentration $N_{h,0} = 10^{16}$ cm^{-3}. Since holes are the majority carriers, this case represents inversion. As the temperature is lowered the thickness of the charge layer is decreased.

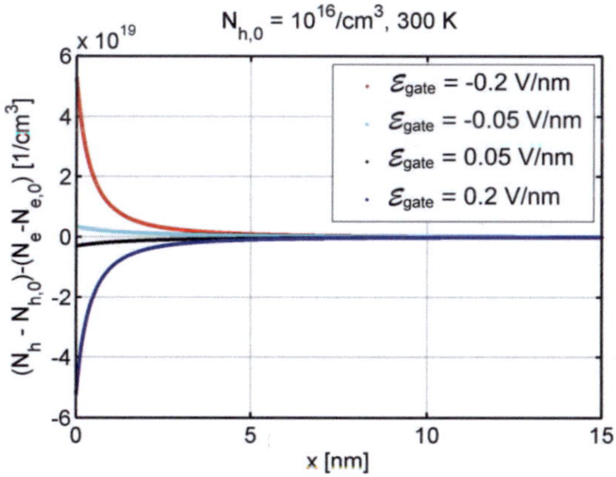

Fig. 7. Accumulation and inversion in p-type silicon. Net carrier density vs. distance for different gate fields (legend) and fixed temperature T = 300 °K.. Silicon is assumed to have a bulk hole concentration $N_{h,0} = 10^{16}$ cm^{-3}. When the net charge is positive accumulation occurs, when the net charge is negative inversion occurs. In both cases the thickness of the charge layer is of few nanometers.

2 High-Speed Silicon-Organic Hybrid (SOH) Modulators

2.1 Introduction

In this chapter we present the results obtained for high-speed silicon-organic hybrid (SOH) Pockels effect modulators.

In Section 2.2 we will present the first SOH modulator performing at a speed of 42.7 Gbit/s. Previously published results were limited to a bandwidth of 3 GHz [4].

In Section 2.3 we show an improved version of the device with higher bandwidth and smaller driving voltage. This device exploits an improved poling scheme of the nonlinear organic cladding. To this end, an appropriate poling chamber was built, and the poling parameters (voltage ramp, temperature, duration) were optimized by extensive experimental testing. Moreover, the novel modulator has a very short length (500 μm), which minimizes the effects of the electrical losses. The result is a modulator having a bandwidth larger than 100 GHz, establishing a new record in silicon photonic modulators.

2.2 42.7 Gbit/s Electro-Optic Modulator in Silicon Technology

L. Alloatti, D. Korn, R. Palmer, D. Hillerkuss, J. Li, A. Barklund, R. Dinu, J. Wieland, M. Fournier , J. Fedeli, H. Yu, W. Bogaerts, P. Dumon, R. Baets, C. Koos, W. Freude, J. Leuthold

Published in Optics Express, Vol. 19, Issue 12, pp. 11841-11851 (2011)

2.2.1 Abstract

CMOS-compatible optical modulators are key components for future silicon-based photonic transceivers. However, achieving low modulation voltage and high speed operation still remains a challenge. As a possible solution, the silicon-organic hybrid (SOH) platform has been proposed. In the SOH approach the optical signal is guided by a silicon waveguide while the electro-optic effect is provided by an organic cladding with a high $\chi^{(2)}$-nonlinearity. In these modulators the optical nonlinear region needs to be connected to the modulating electrical source. This requires electrodes, which are both optically transparent and electrically highly conductive. To this end we introduce a highly conductive electron accumulation layer which is induced by an external DC "gate" voltage. As opposed to doping, the electron mobility is not impaired by impurity scattering. This way we demonstrate for the first time data encoding with an SOH electro-optic modulator. Using a first-generation device at a data-rate of 42.7 Gbit/s, widely open eye diagrams were recorded. The measured frequency response suggests that significantly larger data rates are feasible.

2.2.2 Introduction

One of the most important properties of an optical modulator is its modulation speed or bandwidth, which should be at least as fast as the available electronics. Transistors having transit frequencies well above 100 GHz have already been demonstrated in 90 nm silicon technology, and the speed will further increase as the minimum feature size is reduced [34]. However, the bandwidth of present-day silicon-based modulators is limited. Silicon does not possess a $\chi^{(2)}$-nonlinearity due to its centrosymmetric crystalline structure. The use of strained silicon has led to interesting results already [35], but the nonlinearities do not yet allow for practical devices. State-of-the art silicon photonic modulators therefore still rely on free-carrier dispersion in *pn* or *pin* junctions [18, 36, 37]. Forward biased junctions have been shown to exhibit a voltage-length product as low as $V_\pi L = 0.36$ V mm, but the modulation speed is limited by the dynamics of minority carriers. Still, data rates of 10 Gbit/s have been generated with the help of a pre-emphasis of the electrical signal [37]. Using reverse biased junctions instead, the bandwidth has been increased to about 30 GHz [18, 24], but the voltage-length product rose to $V_\pi L = 40$ V mm. Unfortunately, such plasma effect phase modulators produce undesired intensity modulation as well [38], and they respond nonlinearly to the applied voltage. Advanced modulation formats like QAM require, however, a linear response and pure phase modulation, making the exploitation of the electro-optic effect (Pockels effect [19]) particularly desirable.

2.2.3 SOH Approach

Recently, the silicon-organic hybrid (SOH) approach has been suggested [27, 39-41]. An example of an SOH modulator is shown in Fig. 8(a). It consists of a slot waveguide guiding the optical field, and two silicon strips which electrically connect the optical waveguide to the metallic electrodes. The electrodes are located outside the optical modal field to avoid optical losses [42], Fig. 8(b). The device is coated with an electro-optic organic material which uniformly fills the slot. The modulating voltage is carried by the metallic electrical waveguide and drops off across the slot thanks to the conductive silicon strips. The resulting electric field then changes the index of refraction in the slot through the ultra-fast electro-optic effect. Since the slot has a width in the order of 100 nm, a few volts are enough to generate very strong modulating fields which are in the order of magnitude of the dielectric strength of most materials. The structure has a high modulation efficiency since both the modulating and the optical fields are concentrated inside the slot, Fig. 8(b) [26].

Indeed, first implementations of SOH modulators with sub-volt operation [40] have been already shown, and sinusoidal modulation up to 40 GHz was demonstrated [43, 44]. However, the challenge in building low-voltage high-speed SOH modulators is to create a highly conductive connecting strip. In an equivalent circuit the slot can be represented by a capacitor *C* and the conductive strips by resistors *R*, Fig. 8(b). The corresponding *RC* time constant determines the bandwidth of the device [26, 27, 45, 46]. In order to decrease the resistance *R*, it has been suggested to dope the silicon strips [26, 27]. While doping increases the conductivity of the silicon strips (and therefore increases optical losses), one pays an additional loss penalty because the electron mobility is impaired by impurity scattering [26, 27, 47]. Moreover, the most recent fabrication attempts showed unexpectedly low conductivity [4, 14].

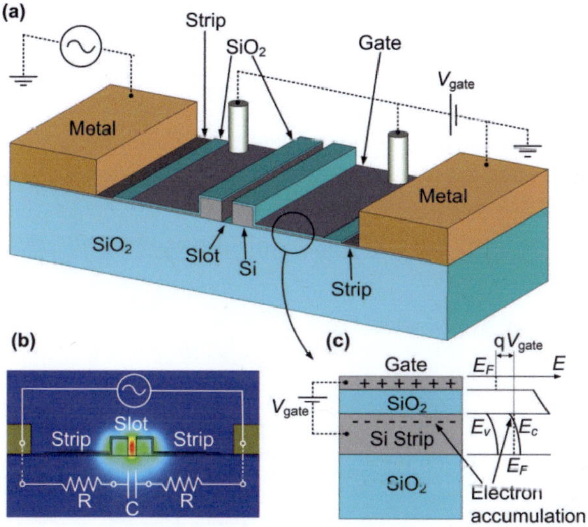

Fig. 8. Vision for a future SOH-modulator. (a) The optical active region is connected to the metal electrodes by means of thin silicon strips. On top of the silicon strips a SiO_2 film is deposited and covered with the gate electrode. (b) Cross-section of the waveguide and electric field distribution of the optical mode; the light is concentrated in the slot. Lower inset: Equivalent RC circuit for the transfer of the voltage between metallic electrodes to the voltage dropping across the slot (slot capacitance C, strip resistance R). (c) When a positive gate voltage is applied across the gate oxide, a highly conductive electron accumulation layer is formed in the silicon strips. Under the effect of the gate voltage V_{gate} the energy bands in the strip are bent. $E_{F,C,V}$ are Fermi energy, conduction and valence band energy, respectively; q is the elementary charge.

2.2.4 The Vision

We present an SOH-based slot-waveguide phase modulator, where the conductivity of the silicon strips connecting the slot region to the modulation electrodes is increased with a novel method which does not create significant optical loss. This enables the first experimental demonstration of data encoding with an SOH modulator. Even though our demonstrator is our very first specimen, which has not at all been optimized, the performance can already compete with that of state-of-the-art plasma-effect modulators [18, 37, 48]. The data rate of 42.7 Gbit/s that we achieved was limited by the available equipment; the modulator performance suggests, however, that significantly higher data rates are possible.

The structure of the modulator we propose is shown in Fig. 8(a). The slot region in the center, where both the optical and the modulating microwave fields interact, is connected to the metal electrodes by thin silicon strips. On top of these strips a slightly conductive layer ("gate") is deposited, isolated from the strips by a thin silicon oxide film. The structure obtained in this way is similar to the well-known metal-insulator-semiconductor (MIS) [33]. When a positive voltage V_{gate} is applied across the oxide (see Fig. 8(a)), the energy bands in the strips are bent (Fig. 8(c)), and a high-mobility electron accumulation layer is formed at the Si/SiO_2 interface. Since the strip conductivity is proportional to the mobility and the free-electron density, the limiting frequency $f_{RC} = 1/(2\pi RC)$ increases with increasing gate voltage. For convenience the structure is referred below as an accumulation-layer electro-optic modulator (ALMod).

The gate of the device depicted in Fig. 8(a) must be optically transparent. This can in principle be achieved by using a thin layer of *n*-doped polysilicon driven to depletion. The ion concentration is adjusted such that when the desired gate voltage is applied all free carriers in the polysilicon are removed. In this way the net positive charge of the gate is mainly caused by fixed nuclei which cannot cause free-carrier absorption.

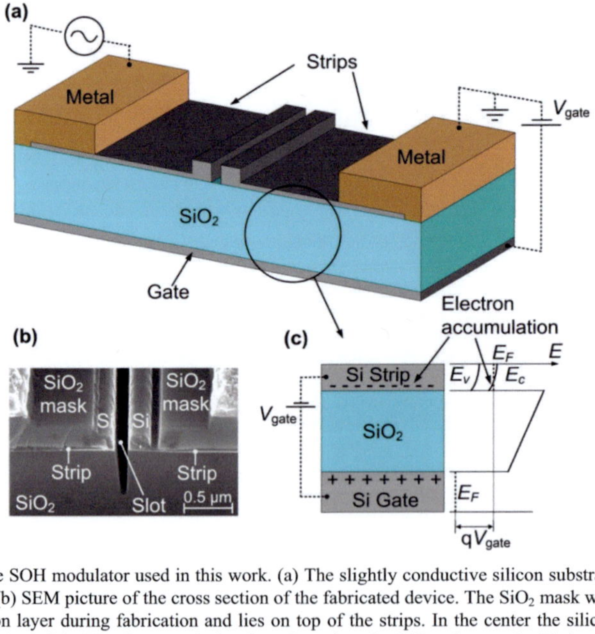

Fig. 9. The SOH modulator used in this work. (a) The slightly conductive silicon substrate is used as a gate. (b) SEM picture of the cross section of the fabricated device. The SiO_2 mask was used as a protection layer during fabrication and lies on top of the strips. In the center the silicon optical waveguide is visible. The slot extends for about 1 µm into the SiO_2 substrate for fabrication issues. (c) When a positive gate voltage is applied across the 2 µm thick SiO_2 substrate a highly conductive electron accumulation layer forms in the strips. The thickness of the strips is 60 nm; for clarity they are not drawn to scale. The gate voltage V_{gate} bends the energy bands in the strips [23]. $E_{F,C,V}$ are Fermi energy, conduction and valence band energy, respectively; q is the elementary charge.

2.2.5 The Fabricated Device

In order to validate the ALMod concept we fabricated the technologically simpler structure depicted in Fig. 9. Here, the gate voltage is applied between the metal electrodes and the silicon substrate. Since the silicon substrate has a conductivity $\sigma = 0.05\ \Omega^{-1}\ cm^{-1}$, the gate voltage drops essentially across the 2 µm silicon oxide which separates the optical layer from the silicon substrate, Fig. 9(a). To avoid any damage to the chip, we did not apply gate voltages larger than 270 V. This corresponds to an electric field $E_{gate} = 0.135$ V/nm in the 2 µm thick SiO_2, which is much smaller than the breakdown voltage of 1 V/nm achievable in state-of-the art thin SiO_2 films [49]. The gate voltage of 270 V can be reduced by more than two orders of magnitude when using few-nanometers-thick gate oxides, as in the optimized structure envisaged in Fig. 8. The device was fabricated in a CMOS fab using deep-UV lithography (see Appendix B), and the commercially available [50] organic material M1 was subsequently deposited and poled *in situ* (see Appendix D). The material M1 consists of chromophores dispersed in amorphous polycarbonate (APC), has a nonlinear electro-optic coefficient up to $r_{33} = 70$ pm/V (see Appendix D) and an index of refraction n = 1.67±0.02, both measured at the wavelength of 1550 nm. The RF dielectric constant is $\varepsilon_r = 3.3 \pm 0.1$ (3.1 ± 0.1) at 10 GHz (60 GHz). A thermal study indicates that operation at 85°C over 25 years results in a change of the nonlinear coefficient by less than 20% [51].

The slot waveguide has a "rail" width and a slot width of 240 nm and 120 nm respectively. The light is coupled by means of grating couplers [52], whose separation is 2.6 mm. The optical waveguide comprises the 1.7 mm long phase shifting section, two tapers and two 67 µm long strip-to-slot waveguide converters [53]. The silicon strips used for connecting electrically the slot waveguide have a thickness of 60 nm (about 10 nm have been oxidized during the high temperature annealing). The ground-signal (GS) metal electrodes have a gap of 4 µm, and their nominal line impedance is 50 Ω. The length of the electrodes is 3.0 mm and comprises two tapered regions for contacting with a length of 0.55 mm each. On the output side, the 50 Ω line is connected to another transmission line which has a length of 3.0 mm and a line impedance of 75 Ω. This transmission line belongs to a second device and is not terminated. It could not be cleaved away since this would lead to an electrical breakdown at the chip edge when applying the gate voltage. The presence of the 75 Ω line is responsible for the local maximum in the S_{21} parameter around 10 GHz as shown in Fig. 11(a); this deviation was not observed in cleaved samples.

In this first prototype the optical losses are high due to a fabrication error (outlined in Appendix B) which led to an increased surface roughness of the waveguides. This resulted in a fiber-to-fiber loss as high as 40 dB, 10 dB of which is due to the grating couplers, 1 dB is due to the two strip-to-slot converters, and 17 dB/mm is due to the rough waveguide. To verify that these limitations come from fabrication issues rather than from fundamental problems, we fabricated test structures and found losses lower than 3.5 dB/mm. While this is already an acceptable value we anticipate fiber-grating losses as little as 1 dB per facet [54] and losses below 1.5 dB/mm based on our previous results [55].

2.2.6 Device Characterization

To demonstrate the high-speed capabilities of the phase modulator we performed a 42.7 Gbit/s data modulation experiment. A 1550 nm laser was used as an optical source, and a pseudorandom bit sequence (PRBS) with a length of $2^{31}-1$ controlled the modulator. A one-bit delay-interferometer (DI) on the receiver side was used to convert phase modulation to intensity modulation, which was then detected with a photodiode. The DI has the further effect of cutting off the low frequency modulation components. The RF voltage swing was set to $V_{pp} = 4.1$ V (measured before the probe), and the device was terminated with an external broadband 50 Ω resistor. When the gate field was increased from zero to its maximum value, the optical loss increased by less than 1 dB. The gate leakage current was always below 10 nA, corresponding to a gate power consumption of less than 3 μW. Clear and open eye diagrams at 42.7 Gbit/s were found with the highest gate field, Fig. 10. Bit-error-ratios (BER) smaller than 3×10^{-10} were recorded, demonstrating the usability of our device in real data links.

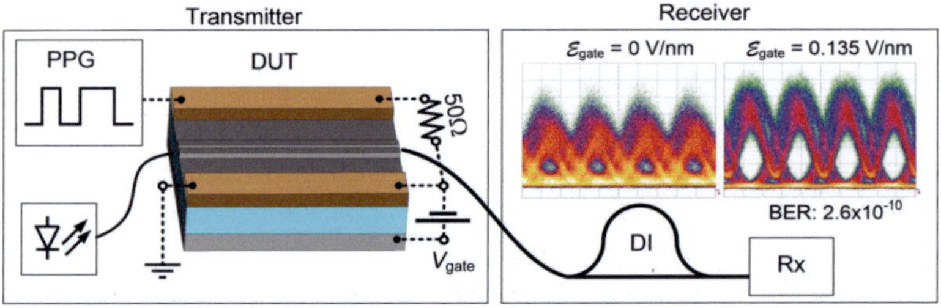

Fig. 10. Setup of the modulation experiment. A pulse pattern generator (PPG) creates a 42.7 Gbit/s electrical signal. Light from a 1550 nm laser is launched into the slot-waveguide. The device is electrically terminated with an external 50 Ω resistor. A gate voltage is applied between the silicon substrate and the silicon strips. The 50 Ω resistance is responsible of keeping both strips at the same electrical DC potential. On the receiver side a delay-interferometer converts the phase modulation into intensity modulation for detection. By increasing the gate voltage to $V_{gate} = 270$ V (gate field $E_{gate} = 0.135$ V/nm) clear and open eyes are found.

We further determined the frequency response of the device by applying a sinusoidal voltage with frequencies f_{mod} between 1 kHz and 60 GHz. The chip was contacted as in the data modulation experiment. The RF power at the probe input was kept constant at 10 dBm (1.0 V amplitude). We measured the resulting phase modulation index η, which represents the achieved phase shift in radians. The data are shown in Fig. 10(a) for different gate fields together with the electrical transmission characteristic of the metal electrodes (S_{21} voltage ratio). By increasing the gate field from $E_{gate} = -0.025$ V/nm to $E_{gate} = 0.135$ V/nm, the modulation index increases by more than a factor of five in the frequency range above 1 GHz. For the highest gate field, a voltage length product of $V_\pi L = 9$ V mm (58 V mm) was measured in the low frequency limit and at 60 GHz, respectively. For our 1.7 mm long device this corresponds to a π-voltage of V_π =5.3 V (34 V). Above 2 GHz, the frequency response is essentially flat (less than 3dB decrease between 2 GHz and 60 GHz) suggesting that data rates could be extended well beyond the 42.7 Gbit/s limit of our equipment.

In order to better investigate the effect of the gate, we recorded the modulation index as a function of the gate field for selected frequencies, Fig. 11(b). For more positive gate fields, an increasing number of electrons accumulate in the silicon strips. The strip resistance decreases accordingly, leading to higher modulation indices. Our simulations indicate that the sheet resistance of the silicon strips becomes smaller than $1500\,\Omega/\text{sq}$ for gate fields $E_{gate} > 0.135$ V/nm , which could otherwise be achieved only by doping the silicon strips with an ion-concentration as high as 3×10^{18} cm^{-3}. At the gate field $E_{gate} = -0.025$ V/nm, a minimum in the modulation index is observed. This indicates that the silicon strips have become highly insulating. Our simulations indicate indeed that for this gate field the slightly *n*-doped strips are fully depleted of free electrons. At more negative gate fields, the modulation index increases again because of the formation of a conductive hole inversion layer.

Charge density and resistance of the real structure were simulated as a function of the gate voltage by means of the two-dimensional simulation package DESSIS [56]. For a qualitative confirmation, both quantities were then compared with the numerical solution of the one-dimensional MIS equations [33]. Both methods predicted the three different regimes (accumulation, depletion and inversion), as observed in our experiments.

2.2.7 Discussion and Conclusion

In this first implementation, the poling of the organic cladding is incomplete. By exploiting the full potential of organic electro-optic materials, the drive voltage can be decreased by a factor of eight (see Appendix D). Moreover, the electrical performance of the current metal electrodes is not optimal, Fig. 11(a). This can be improved in future devices by increasing the thickness of the metallization [26], or by using a distributed on-chip RF driver [57].

The ALMod structure proposed in Fig. 8 offers a number of advantages when compared with structures made conductive by ion implant [26, 27]. First, the silicon strips of Fig. 8 can be made notably thinner than the 60 nm used in Fig. 9. In fact, thin strips are good enough since for gate fields of 0.135 V/nm or higher, more than 90 % of the free carriers are already concentrated in the first 10 nm from the Si/SiO_2 interface. Also, for thin strips the optical field is concentrated more strongly inside the slot leading to a more efficient modulation [26]. The second advantage is that the strips do not require doping anymore, so that the electron mobility remains unperturbed and high [58], leading to lower optical losses for a given conductivity (see Appendix A). Third, when using thin gate oxides (in the order of 10 nm), gate fields seven times larger than those used in our work can be applied [49] decreasing the strip resistances even more. This will become important once narrower slots are fabricated, since then an additional factor ten in the modulation efficiency can be gained [26], at the price of a larger slot capacitance.

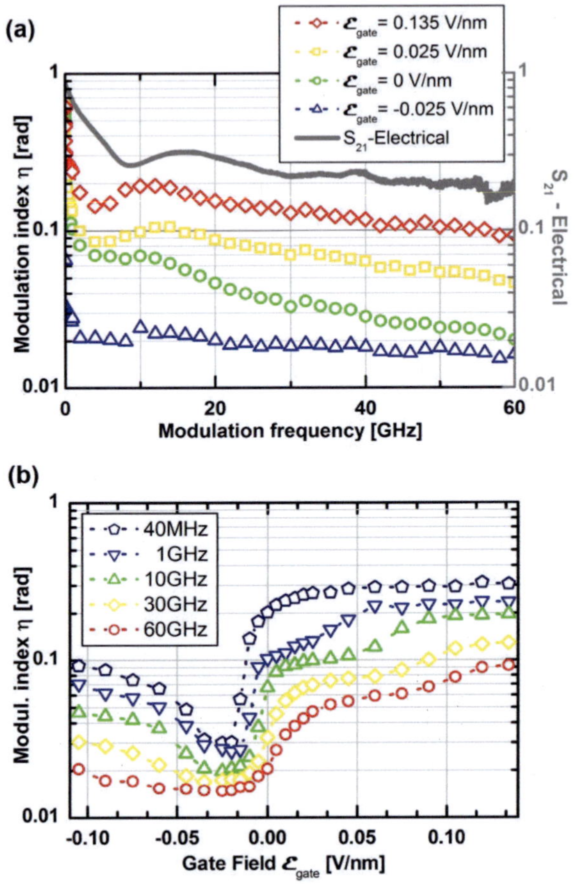

Fig. 11. Response of the DUT vs. frequency and gate field for 1 V modulation amplitude. (a) Phase modulation index η vs. frequency for selected electrical gate fields. The measured modulation frequency range is 1 kHz to 60 GHz. When varying the gate field from –0.025 V/nm to 0.135 V/nm the silicon strips become more conductive, and the modulation index increases accordingly. The gray curve is the S_{21} electrical transmission of the metallic electrical waveguide (voltage ratio); the decrease to 0.3 is an undesired effect of the electrical losses of our test device. (b) Modulation index vs. gate field for selected modulation frequencies. Each curve reaches a plateau at high gate fields. For a gate field strength of $E_{gate} = -0.025$ V/nm the silicon strips become highly insulating, therefore a minimum is observed. At more negative fields the modulation increases again because of the formation of a conductive hole inversion layer in the silicon strips.

In summary, we demonstrated the first 42.7 Gbit/s operation of an SOH electro-optic modulator. We introduced a method for increasing the conductivity of the thin silicon electrodes by using a novel electron accumulation-layer technique. This way, the modulation increased by a factor of five at 60 GHz while the optical loss increased by less than 1 dB.

2.2.8 Appendix

A. Optical absorption caused by injected carriers

The well-known empirical equations of Soref and Bennett [59] relating the optical absorption in silicon to the free carrier density is based on experimental data where the free carriers come from impurity ionization and not from carrier injection. Soref and Bennett explicitly assumed equivalence between the two cases [59]. Instead, the Drude-Lorenz model predicts that the optical absorption coefficient is proportional to the free carrier density N and to the inverse of the mobility μ. In doped samples the mobility is strongly impaired by scattering at impurities, while this is not the case for injected carriers in pure silicon [58]. As an example, the electron mobility drops from roughly 1200 $cm^2/(Vs)$ to less than 300 $cm^2/(Vs)$ passing from an impurity concentration of 10^{16} cm^{-3} to 10^{18} cm^{-3} [60]. As a consequence, for a given free electron concentration N_e, not only the conductivity $\sigma = q\,N_e\,\mu_e$ is lower for doping than for injection, but also the optical loss , $\alpha_{\mathrm{Drude}} \propto N_e/\mu_e$ is higher. This indicates that injected carriers have an intrinsic advantage over carriers coming from impurity ionization when both, high conductivity and low optical loss, are required [61].

B. Waveguide fabrication

The waveguides were fabricated within the ePIXfab framework by CEA-LETI according to standard processes of the microelectronic industry [62]. Two silicon-on-insulator (SOI) wafers were used, with layers of 220 nm thick crystalline silicon and 2 µm thick silicon oxide. The wafers were first implanted with arsenic for reaching a uniform ion concentration of $10^{17}/cm^{3}$. To this end, a dose of $5 \times 10^{12}/cm^2$, energy of 150 keV, tilt 0°, and twist 0° was used. Grating couplers [52] were then created by means of a 248 nm DUV lithography followed by 70 nm silicon etch performed with HBr. A high temperature oxide (HTO) layer of 130 nm used as a hard mask was grown and subsequently structured by means of 193 nm DUV lithography. A 150 nm silicon etch was used to define the waveguides. Taking advantage of the self-alignment with the pre-existing hard-mask, a 248 nm lithography defined the regions for a full silicon etch inside the slots. A thermal oxidation at 1100°C for 10 minutes was performed for reducing the surface roughness. A 248 nm DUV lithography defined the high-doping regions for the formation of ohmic contacts beneath the metal electrodes. To this end, the wafers were implanted with arsenic using a dose of $2 \times 10^{15}/cm^2$ and an energy of 30 keV. The ions were activated with a 1050°C annealing for 15 minutes. At this point, one wafer was not processed further for future optical loss measurements. Waveguides belonging to this wafer, but also coated with the nonlinear organic material, had an insertion loss of 16 dB (10 dB of which is caused by the grating couplers). A 500 nm thick silica protection layer was deposited on the entire wafer. A 248 nm DUV lithography was followed by a silica etch down to the silicon layer. A Ti/TiN/AlCu metal stack having a total thickness of 600 nm was deposited by physical vapor deposition (PVD) on the entire wafer. A 248 nm DUV lithography and a reactive ion etch (RIE) with chlorine was used to structure the metal electrodes. The wafer was then annealed for 30 minutes at 425°C. Finally, the slots were opened by means of a silica etch. The waveguides were spin-coated with the nonlinear organic material and poled *in situ*. The insertion loss of the waveguide is 40 dB. This 24 dB increase is mostly due to an error that occurred in the last silica etch step, which considerably increased the surface roughness of the waveguide. We are currently working on improving this process.

C. Method used for determining the modulation index

In the frequency range between 40 MHz and 60 GHz we derived the phase modulation index η by evaluating the ratio $J_0^2(\eta)/J_1^2(\eta)$ between the central intensity peak and the first sideband intensity of the phase modulated signal (J_ν is the Bessel function of the first kind). The spectra were recorded with an optical spectrum analyzer (Apex AP2050). In the frequency range between 1 kHz and 40 MHz we inserted the phase modulator in one arm of a fiber-based Mach-Zehnder interferometer. The interference was recorded by means of a wide-band photodetector (Thorlabs PDA10CF) and a 1 GHz oscilloscope. The achieved phase modulation was derived from the amplitude of the intensity modulation.

D. Poling of the electro-optic material and origin of the phase shift.

The commercially available nonlinear material M1 [50] was poled by applying 16 V to the metal electrodes while the device was heated from room temperature to 141°C and then rapidly cooled as soon as this temperature was reached. Unpoled samples showed no detectable spectral sidebands (phase modulation smaller than 0.002 rad/V), demonstrating that the measured phase shift is actually due to the cladding nonlinearity and is not caused by free carriers. From the modulation index in the low frequency limit we estimate an actual nonlinearity coefficient of $r_{33} = 20 \pm 2$ pm/V. There is large potential for further increasing this value: The same material is used in commercially available polymer modulators where values of $r_{33} = 70$ pm/V are routinely achieved by parallel plate poling. In SOH systems, values as high as $r_{33} = 40$ pm/V have been reported [4]. The highest r_{33} value achieved by *in situ* poling amounts to 170 pm/V [22]. This would result in a reduction of the driving voltage by a factor of eight.

2.3 100 GHz Silicon-Organic Hybrid Modulator

L.. Alloatti, R. Palmer, S. Diebold, P. Pahl, M. Pauli, J. Li, B. Chen, R. Dinu, M. Fournier, J. Fedeli, T. Zwick, C. Koos, W. Freude, J. Leuthold

Submitted for publication

2.3.1 Abstract

We demonstrate for the first time an integrated silicon modulator with a 3 dB bandwidth of 100 GHz. The second-order nonlinearity is contributed by a highly-nonlinear organic cladding. Despite the high speed of our device, its voltage length product is only $V_{\pi}L$ = 11 Vmm, which very well competes with the best silicon-photonic modulators available today.

2.3.2 Introduction

High-bandwidth electro-optic modulators are key components for a variety of applications such as photonic transceivers for long-haul and on-chip communications [63], radio-over-fiber links, low-noise microwave oscillators [64] or optical frequency comb generation [65]. However, achieving small power consumption, low modulation voltage, and high-speed operation still remains a challenge. Since unstrained silicon does not possess a $\chi^{(2)}$-nonlinearity [35], state-of-the art silicon photonic modulators mainly rely on free-carrier dispersion (plasma effect) in *pin* or *pn* junctions [18, 37, 66]. Reversed-biased *pn* junctions are intrinsically faster than forward biased *pin* diodes [37] and enabled already 50 Gbit/s on-off keying (OOK), with a voltage-length product $V_{\pi}L$ = 28 V mm [67]. Unfortunately, such plasma effect phase modulators produce undesired intensity modulation as well, and they respond nonlinearly to the applied voltage. An alternative approach uses hybrid integration of III-V epitaxy stacks grown on InP substrates, which are subsequently transferred to SOI waveguides for creating high-speed electro-absorption modulators [68]. Recently, such a device demonstrated a 3 dB bandwidth larger than 67 GHz, representing so far the fastest modulator realized on a silicon chip. Advanced modulation formats like QAM, however, require phase modulators with a linear response and a pure phase modulation, rendering the electro-optic effect (Pockels effect [19]) particularly attractive. For small power-consumption, devices with small capacitances are required, and leakage currents as well as driving voltages are to be kept minimum [26].

In this paper we demonstrate for the first time an electro-optic modulator having a 3 dB bandwidth larger than 100 GHz. The device has a voltage-length product of $V_{\pi}L$ = 11 V mm – the smallest ever reported value for a high-speed silicon-based modulator. Our modulator exploits the linear Pockels effect, offers capacitances per device length as small as 100 fF/mm [26], and a parasitic resistance of several GΩ.

2.3.3 SOH modulator

Our device is based on the silicon-organic hybrid (SOH) technology. In an SOH modulator, the optical quasi-TE field (dominant electric field parallel to the substrate plane) is guided by a silicon waveguide, while the electro-optic effect is provided by an organic cladding with a high $\chi^{(2)}$-nonlinearity (Pockels effect), Fig. 12(a). The optical nonlinear interaction occurs inside a nanoscale slot, taking advantage of the field-enhancement caused by the lateral discontinuity of the refractive index, Fig. 12(b). The external modulation voltage drops across the nanoscale slot thanks to strip-load silicon electrodes which must be both optically transparent and electrically highly conductive, Fig. 12(a).

To reduce the resistance of the silicon strip-loads, we exploit a highly-conductive electron accumulation layer which is induced by an external DC "gate" voltage, Fig. 12(c). In contrast to doping, the electron mobility is not impaired by impurity scattering so that sheet resistances as low as 3 kΩ/sq can be obtained together with small optical loss [3]. This is a key technique for large modulation bandwidths, and helps overcoming the *RC* speed limitation so that the device speed is limited only by the frequency-dependent RF propagation losses [46].

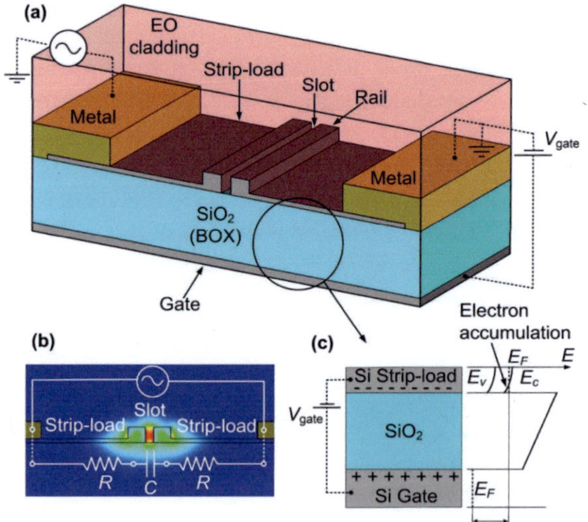

Fig. 12. Schematic of an SOH modulator. (a) The silicon slot waveguide is connected to the metal electrodes by thin silicon strip-loads. The electro-optic (EO) polymer cladding covers the waveguide and fills the slot. The shallowly doped silicon substrate is used as a gate. (b) Cross-section of the waveguide and electric field distribution of the optical quasi-TE mode; the light is concentrated in the slot. The figure also shows the equivalent *RC* circuit for the modulation voltage (slot capacitance *C*, strip-load resistance *R*). (c) When a positive gate voltage is applied across the 2 µm thick SiO_2 substrate a highly conductive electron accumulation layer forms in the strip-loads. The thickness of the strip-loads is 60 nm (not drawn to scale). The gate voltage V_{gate} bends the energy bands in the strip-loads [33]. $E_{F,C,V}$ are Fermi energy, conduction and valence band energy, respectively; q is the elementary charge.

2.3.4 Device fabrication and characterization

The optical waveguide has been fabricated in a CMOS fab according to standard processes of the microelectronics industry (see Appendix). The width of the slot is 120 nm, and the rail width is 240 nm. The buried-oxide (BOX) has a thickness $d_{BOX} = 2$ µm. We used silver electrodes with a length of only 500 µm to overcome the speed limitations caused by the frequency-dependent RF propagation losses, which were predicted to be as high as 4.8 ±0.8 dB/mm @ 65 GHz and 6.1 ±1 dB/mm @ 100 GHz [26] for a similar device,. The chip was coated with the commercially available nonlinear polymer M3 [50], and poled in-situ [3, 15] thereby obtaining an electro-optic coefficient of $r_{33} = 18$ pm/V. This material has been successfully tested by Telcordia [69].

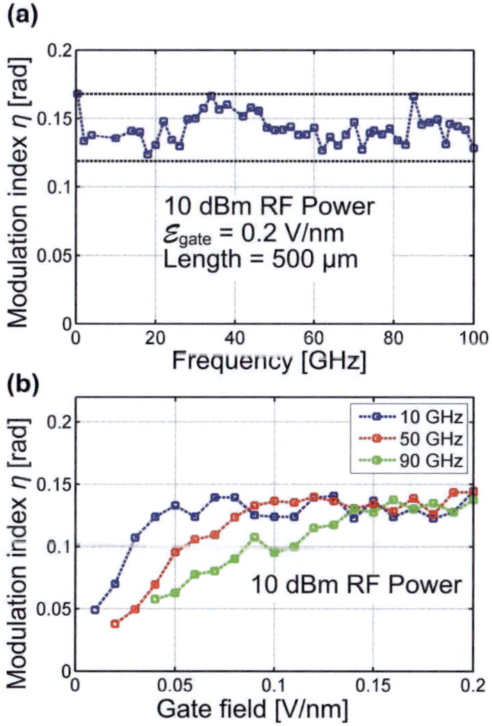

Fig. 13. Modulation index η vs. frequency and gate field, respectively. The data are normalized to 10 dBm launched power (1 V amplitude at 50 Ω characteristic impedance). (a) Phase modulation index η vs. frequency for a gate field of 0.2 V/nm. The measured modulation frequency range is 0.4 GHz to 100 GHz. The horizontal dotted black lines represent the maximum value, and 70.7 % of the maximum value which corresponds to the 3 dB bandwidth of at least 100 GHz (b) Modulation index vs. gate field for selected modulation frequencies. Each curve reaches a plateau at high gate fields indicating that in this region the charging of the slot capacitance is not limited by the strip-load resistance.

For determining the frequency response of the device, we applied a sinusoidal signal with frequencies f_{mod} between 0.4 GHz and 100 GHz. We used three independent RF sources and calibrated electrical power meters. The chip was contacted with two Picoprobes (GGB Industries Inc, Models 67A GS/SG 50 P) with a 50 µm pitch. The chip was terminated with an external 50 Ω load. The losses of the RF cables and Picoprobes were subtracted using a ceramic calibration substrate (GGB, CS-8). The modulation index η (i.e. the amplitude of the sinusoidally varying phase shift) was extracted from the relative height of the sidebands recorded with an optical spectrum analyzer (Apex AP2050) [3]. After checking the linearity of the modulation index with respect to the applied modulation voltage, the modulation indices were rescaled to a reference 10 dBm RF power (1 V amplitude at 50 Ω impedance). The frequency response of the modulator varies by less than 30 % in a 100 GHz frequency range, meaning that if the device was mounted in one arm of a Mach-Zehnder interferometer (MZI) the 3 dB bandwidth of the modulator would be at least 100 GHz; Fig. 13(a).

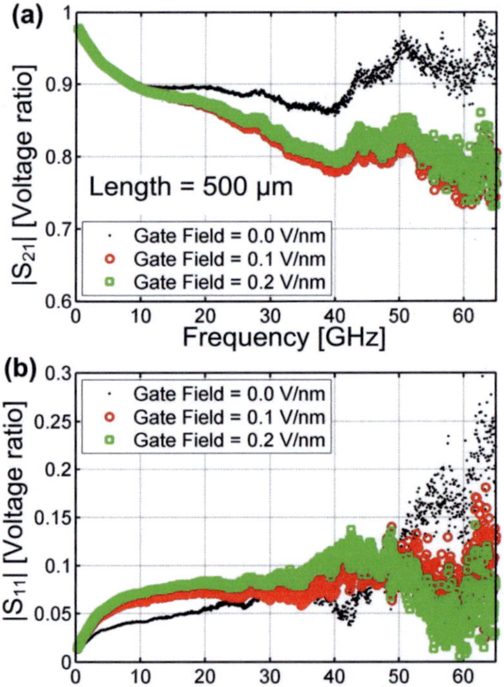

Fig. 14. Electrical S-parameters of the modulator. (a) Electrical transmission factor $|S_{21}|$ (voltage ratio). As the gate field is increased from 0 V/nm to 0.1 V/nm, the electrical loss increases at frequencies > 10 GHz due to the increasing influence of the strip-loads [26]. For even higher gate fields the transmission is almost independent of the gate voltage and equals about $|S_{21}|$= 0.8 at 65 GHz indicating that the RF is not unduly attenuated. (b) Electrical reflection factor $|S_{11}|$ (voltage ratio) for different gate fields.

As we shall show, the *RC* time constant does not limit the device bandwidth. To this end, we monitored the modulation index as a function of the gate field $E_{gate} = V_{gate}/d_{BOX}$ for three different modulation frequencies, Fig. 13(b). As the gate field is increased, the resistance of the strip-loads is diminished due to the formation of the highly-conductive accumulation layer. The modulation index reaches a plateau even at 90 GHz, indicating that charging the slot capacitance is not impeded by the strip-load resistance, Fig. 13(b).

Next, we investigated the performance of the 500 μm long silver electrodes using a vector network analyzer (VNA, Anritsu 37397C) which operates in the range from 40 MHz to 65 GHz, Fig. 14. The system was calibrated such that the reference planes are located on the tips of the Picoprobes. The scattering matrix element S_{21} is defined to be the ratio of the voltages of the outgoing wave on the 50 Ω Picoprobe and the incoming wave; the reflection factor S_{11} is defined as the voltage ratio of incoming and outgoing wave at the same port. As the gate field is increased, the strip-loads become sufficiently conductive for quickly charging the slot capacitance, leading to higher RF losses (decreasing transmission $|S_{21}|$) [26]. After travelling along the 500 μm long electrodes, the forward propagating wave at 65 GHz has still 80 % of the initial voltage amplitude. The spatially averaged modulation voltage is therefore close to 90 %, consistent with the measured optical response. The characteristic impedance increases slightly with frequency [26]. Based on our VNA measurements, we extract a characteristic impedance of 40 Ω @ 10 GHz and 50 Ω @ 60 GHz for the maximum gate field. The slight oscillation visible both in the electrical (Fig. 14) and in the optical response (Fig. 13(a)) in the vicinity of the modulation frequency of 40 GHz is attributed to the slight impedance mismatch. With an earlier device generation, having similar π voltage but smaller bandwidth, we already showed 42.7 Gbit/s data transmission with a bit-error-ratio (BER) smaller than 3×10^{-10} [3]. The current device is expected to support on-off-keying (OOK) data rates up to 140 Gbit/s.

2.3.5 Discussion and conclusion

We demonstrated the first silicon hybrid phase modulator having a 3 dB bandwidth of at least 100 GHz. With a gate voltage we induce an accumulation layer for reducing the resistance of the connecting silicon region between electrodes and optically active region. This reduces the RC time constant to a point where it does not limit the device bandwidth. Our device is very short (500 μm) and has a voltage-length product as small as $V_\pi L = 11$ V mm, a value which very well competes with state-of-the-art high-speed silicon modulators. This value can be further improved by an order of magnitude by using materials with stronger nonlinearity [21].

2.3.6 Appendix

The optical waveguides were fabricated by CEA-LETI according to standard processes of the microelectronic industry [62]. Silicon-on-insulator (SOI) wafers were used, with 220 nm crystalline silicon on top of 2 μm silicon dioxide. The wafers were first *n*-doped (arsenic) for reaching a uniform ion concentration of $10^{17}/\text{cm}^{-3}$. To this end, a dose of $5 \times 10^{12}/\text{cm}^2$, an energy of 150 keV, a tilt of 0°, and a twist of 0° were used. Grating couplers [52] were then created by 248 nm DUV lithography followed by a shallow 70 nm silicon etch performed with HBr gases. A high temperature oxide (HTO) layer of 130 nm used as a hard mask was deposited and subsequently structured by 193 nm DUV lithography. A 150 nm silicon etch was used to define the waveguides. Taking advantage of the self-alignment with the pre-existing hard-mask, a 248 nm lithography defined the regions for a full silicon etch inside the slots. A thermal oxidation at 1100°C for 10 minutes reduced the surface roughness. A 248 nm DUV lithography defined the highly-doped regions for the formation of ohmic contacts beneath the metal electrodes. To this end, the wafers were implanted with arsenic using a dose of $2 \times 10^{15}/\text{cm}^2$ and an energy of 30 keV. The ions were activated with a 1050°C annealing for 15 minutes. After defining the resist mask for the electrodes, the chips were dipped for 15 s in buffered hydrofluoric acid (BOF) for removing the oxide. A Cr/Ag metal stack was deposited by e-beam PVD with a pressure of 10^{-6} mBar (5 nm of chromium, 400 nm of silver) and subsequently processed by lift-off. The optically nonlinear cladding was spin-coated at 3000 rpm for 30 s and poled at 167 °C.

3 Second-Order Nonlinear Silicon-Organic Hybrid (SOH) Waveguide

3.1 Abstract

We describe a concept for second-order nonlinear optical processes in silicon photonics. A silicon-organic hybrid (SOH) double slot waveguide is dispersion-engineered for mode phase-matching (MPM). The proposed waveguide enables highly efficient nonlinear processes in the mid-IR range. With a cladding nonlinearity of $\chi^{(2)} = 230$ pm/V and 20 dBm pump power at a CW wavelength of 1550 nm, we predict a gain of 14.7 dB/cm for a 3100 nm signal. The suggested structure enables for the first time efficient second-order nonlinear optical mixing in silicon photonics with standard technology.

This chapter is based on the following paper:

Second-Order Nonlinear Silicon-Organic Hybrid Waveguides

L. Alloatti, D. Korn, C. Weimann, C. Koos, W. Freude and J. Leuthold

Published in Optics Express, Vol. 20, Issue 18, pp. 20506-20515 (2012)

3.2 Introduction

Second-order nonlinear processes, like sum- and difference-frequency generation, spontaneous down-conversion and optical parametric amplification [19], are essential for a number of applications, ranging from spectroscopy, free-space communication, biochemical sensing, medical therapy [70], ultra-fast optical signal processing [71], lowest-noise optical amplification [72], and quantum physics [73]. Since at least one of the frequencies involved in a three-wave mixing process is necessarily well separated from the others, second-order processes represent additionally an excellent candidate for generating mid-IR and far-IR wavelengths [74].

Efficient nonlinear conversions require materials with strong nonlinearities, high-optical intensities and phase-matching between the waves involved [19]. So far, second-order nonlinear waveguides have been made of polymers, GaAs, InP, and $LiNbO_3$ [75], and phase-matching has successfully been achieved by birefringence, intermodal dispersion, or quasi-phase-matching. These materials often require specialized process technologies which are not applicable for mass production. Further, the relatively small index contrasts which can be achieved limit the modal confinement in waveguides. Conversion efficiencies are hence small and optical output powers in the μW range are considered high [75].

Silicon photonics, on the other hand, is based on a widely available technology, and allows fabricating high-index contrast waveguides for obtaining the required intensities with low optical powers. The vision of creating mid-IR applications using the inexpensive silicon-photonics platform [76] has already led to a number of publications in the following topics: Silicon waveguides pumped below the two-photon absorption (TPA) edge with powers as high as 33.5 W (45 dBm) [77], low-loss propagation in the 2-6 μm wavelength range [78, 79], light generation at 2.4 μm with standard telecom sources [80], high-Q SOI photonic crystal cavities at 4.4 μm [81], Raman amplification at 3.39 μm [82], and extensive simulations for single-mode operation and polarization-independent operation in SOI rib waveguides in the mid-IR region [83].

However, unstrained crystalline silicon is centrosymmetric, and its second-order nonlinearity is vanishing [35]. As a consequence, mid-IR generation in unstrained silicon waveguides has to rely on the third-order nonlinearity [77], taking advantage of the "built-in" strong Kerr nonlinearity of crystalline [84] or amorphous silicon [85]. This, in turn, results in high pump power requirements.

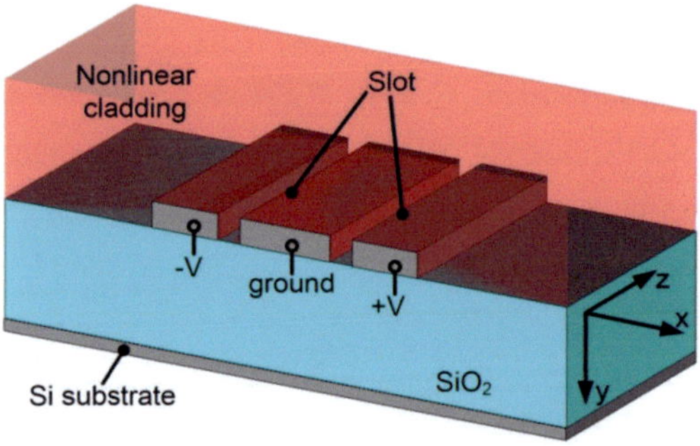

Fig. 15. Silicon organic-hybrid (SOH) double slot waveguide for second-order nonlinear applications. The waveguide consists of three silicon strips on a glass substrate, it is multimode and dimensioned such that modal phase-matching is achieved. The waveguide is covered by a nonlinear cladding, which is poled during fabrication by applying the voltages –V and +V to the outermost strips while the central strip is grounded. As a result the nonlinear second-order susceptibility is high only inside the slots.

In silicon photonics, first attempts of second-harmonic generation are based on strained waveguides, but despite nanosecond peak pump powers as high as 0.7 W, the output peak powers are limited to only 40 nW [86]. The small efficiency is due to a non-phase-matched design. A method for achieving quasi-phase-matching (QPM) in periodically strained silicon has already been proposed [87], however large mode sizes and small nonlinearities lead to normalized conversion efficiencies smaller than 1 % $W^{-1}cm^{-2}$. More recently, a method of achieving phase-matching based on birefringence in strained silicon waveguides has been proposed [88], but the efficiency of the device relies on nonlinearities which have not been shown so far in waveguides of the proposed size [89]. Finally, the potential high efficiencies of SOH second-order nonlinear waveguides have already been discussed [39], but unfortunately no waveguide design has been proposed so far.

In this paper we propose for the first time a second-order nonlinear silicon-organic hybrid (SOH) waveguide [46] based on standard silicon-on-insulator (SOI) technology. The device is suited for efficient mid-IR generation and amplification, Fig. 15. We predict optical pump powers that are orders of magnitude smaller than for Kerr-nonlinear devices. The nonlinearity is provided by a specially engineered nonlinear organic cladding [22, 90], and a novel double slot geometry enables both strong interaction with the nonlinear material and phase-matching. We confirm the viability of our concept by extensive numerical simulations and give design guidelines for covering a broad wavelength range.

This paper is structured as follows: In Section 2 we describe the structure of the proposed device. In Section 3 the phase-matching condition is investigated in detail. In Section 4 the conversion efficiency is calculated and the required optical powers are discussed. In Appendix A, we present a mode-converter for exciting the required modes. In Appendix B we compare the optical electric field strengths needed for generating a given nonlinear polarization. We show that the required pump field is orders of magnitude smaller for second-order than for third-order nonlinearities.

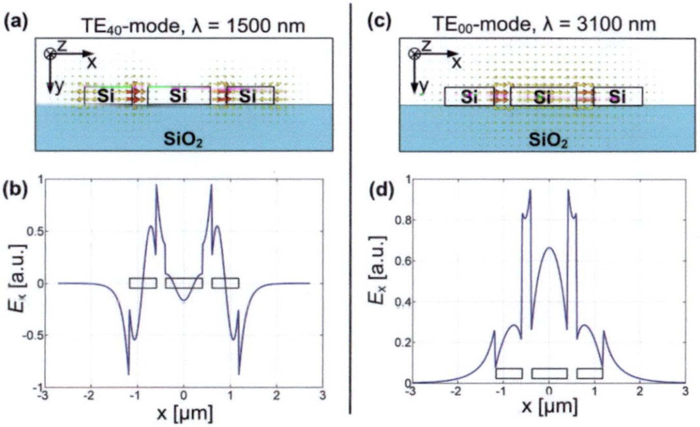

Fig. 16. Transverse electric fields for pump mode and signal/idler mode. (a) Quasi-TE_{40} pump mode at $\lambda = 1500$ nm. Light is strongly confined to the two slots where the nonlinear interaction takes place. (b) Dominant horizontal electric field component E_x of TE_{40} mode at half the strip waveguide height. (c) Fundamental quasi-TE_{00} mode at $\lambda = 3100$ nm for signal and idler. Also in this case, light is strongly confined to the two slots. (d) Dominant horizontal electric field component E_x of the fundamental TE_{00} mode at half the strip waveguide height.

3.3 The Device Concept

A sketch of the proposed second-order nonlinear device is displayed in Fig. 15. It consists of three parallel silicon strips (double slot waveguide) realized on a standard silicon-on-insulator

(SOI) wafer having an oxide thickness of 2 µm and a device layer of 220 nm. The waveguide is spin coated with polymer-dispersed nonlinear chromophores [50], which have a high $\chi^{(2)}$-nonlinearity only inside the two slots. This can be experimentally achieved by poling [3] the material at its glass transition temperature, and by applying two voltages as shown in Fig. 15.

For definitiveness, we will describe a device for difference-frequency generation (DFG) with a pump wavelength of 1.5 µm, a signal wavelength of 2.9 µm and an idler wavelength close to 3.1 µm. Signal and idler propagate in a mode different from the mode in which the pump light is guided. The concept, however, is more general and may be applied to other spectral ranges as will be outlined below.

The waveguide is dimensioned such that modal phase-matching (MPM) is achieved [75] between the quasi-TE_{40} mode for the pump (four nodes in the horizontal direction, zero nodes in the vertical direction) and the fundamental quasi-TE_{00} mode for signal and idler; the corresponding mode profiles are shown in Fig. 16. These modes can be excited efficiently with mode converters described in Appendix A. We did not find any combination of (the first six) modes which could satisfy MPM in waveguides with a single slot, and the solution presented here is the one with the lowest mode-order that we could find in double slot waveguides. Finally, it is worth noticing that dispersion engineering in double slot waveguides has already been exploited in the context of third-order nonlinear processes [91].

3.4 Phase-Matching

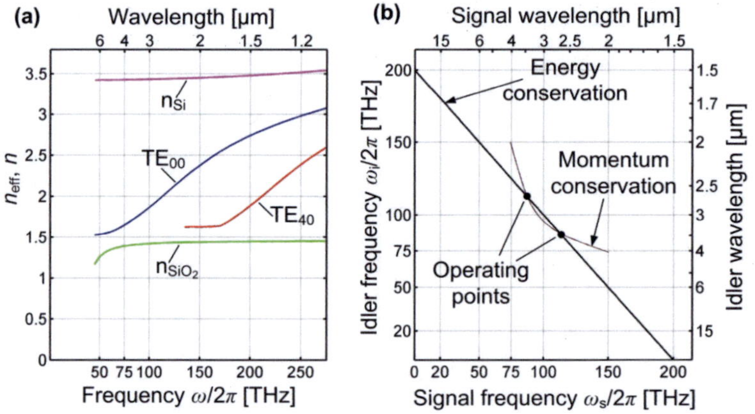

Fig. 17. Mode dispersion and phase-matching conditions for a typical waveguide. (a) Effective index n_{eff} for TE_{00} and TE_{40} TE-modes are plotted vs. frequency (bottom axis) or wavelength (top axis). The refractive indices of silicon and of silicon dioxide used in the simulation are also plotted. (b) Signal and idler frequencies for energy and momentum conservation. The straight line describes energy conservation while he curved line represents the condition for momentum conservation according to Eq. (3.1). Geometry considered in this example: Side-strip/slot/central-strip widths are 580/200/800 nm. In order to take into account the material dispersion of the nonlinear polymer cladding, a refractive index of 1.58 was assumed for calculating the signal and idler mode, while a refractive index of 1.68 was assumed for the pump mode.

The frequencies at which MPM is achieved depend on the dispersion of the (quasi-) TE_{00} and the (quasi-)TE_{40} modes. As an example, we will now consider a waveguide dimensioned as follows: Width of the outermost strips 580 nm, slot width 200 nm, width of the central strip 800 nm; these values are well within the capabilities of current silicon-photonic foundries. The dispersion diagram of the TE_{00} and the TE_{40} modes is shown in the frequency range from 50 THz to 250 THz (wavelength range from 6 μm to 1.2 μm), Fig. 17(a). Material dispersion for modeling the refractive index of the thermal oxide beneath the silicon waveguide is taken from [92], while the corresponding data for crystalline silicon are from [93]. The corresponding curves are also shown in Fig. 17(a).

In Fig. 17(b) we analyze the conditions for MPM, based on the data of Fig. 17(a). The straight black line represents the dependence of idler frequency on signal frequency which have to sum up to the pump frequency (in this example 200 THz, energy conservation), and the curved line represents the points where phase-matching is achieved (momentum conservation) [19],

$$\begin{aligned} &\hbar\omega_s + \hbar\omega_i = \hbar\omega_p, \\ &\hbar k_s + \hbar k_i = \hbar k_p, \qquad k_{p,s,i,} = n_{\text{eff},\,p,s,i}\,\omega_{p,s,i} / c, \end{aligned} \tag{3.1}$$

where c is the speed of light, k is the wavevector component along the waveguide direction, ω is the angular frequency and s, i, p stand for signal, idler and pump respectively. The intersections of the two curves determine the operating points of the device, i. e., they fix signal and idler frequencies. As can be seen from Fig. 17(b), a pump frequency of 200 THz yields operating points with signal and idler frequencies at about 115 THz and 85 THz.

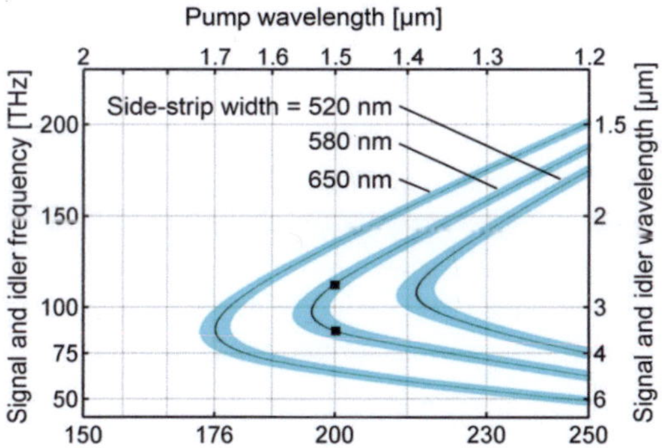

Fig. 18. Signal and idler frequencies vs. pump frequency for three different geometries. The black curves specify for a given pump frequency the signal and idler frequencies which satisfy the energy conservation and the phase-matching condition Eq. (3.1). The cyan-coloured regions indicate the frequency space where the coherent buildup length $L_{\text{coh}} = 2 / (k_s + k_i - k_p)$ [19] is equal to 1 cm or longer. The three different curves represent waveguides where the side-strip width is set to 520 nm, 580 nm and 650 nm. The central-strip width is 800 nm and the slot width is 200 nm in all the three cases. For a side-strip width of e.g. 580 nm and a pump wavelength of 1.5 μm (200 THz), signal and idler wavelengths of 2.6 μm and 3.5 μm would result (square symbols).

Choosing a different pump frequency will result in other operating points that strongly depend on the particular dispersion of the waveguide. This information is represented in Fig. 18. The three curves stand for three different waveguide geometries which can be chosen by the user according to the desired pump and signal wavelengths. The cyan regions show the frequencies for which the coherent buildup length $L_{\text{coh}} = 2 / (k_s + k_i - k_p)$ [19] is equal or larger than 1 cm. A wavelength detuning of 50 nm or more from the ideal therefore still allows a coherent buildup of the converted wave, showing that the wavelength tuning requirements are very relaxed. From Fig. 18 it can also be deduced that the required fabrication tolerances are in the 10 nm range, i. e. within the capabilities of today's silicon photonics foundries. Additionally, the strip and slot size are relatively large [3], meaning that the waveguide can be fabricated by means of standard 193 nm DUV lithography [94]. The full fabrication of the device consists therefore in creating the passive structures in a CMOS fab followed by a post-processing spin coating and poling of the nonlinear material.

3.5 Power Levels and Conversion Efficiencies

In this section we analyze numerically the performance of the proposed waveguide and show that small optical power levels can lead to significant optical output power or optical gain. To this end, we consider here again the case of difference-frequency generation (DFG); a similar analysis can be applied to any other three-wave mixing process.

We define the nth complex electric field component ($n = 1, 2, 3$) in a Cartesian coordinate (x, y, z)- system by separating a dimensionless amplitude $A(\omega)$ from the modal field $\mathcal{E}_n^+(\omega, x, y)$. A propagator describes the wave propagation of a monochromatic wave with angular frequency ω and propagation constant k along the z-direction,

$$E_n(x, y, z, t) = A(\omega, z)\mathcal{E}_n^+(\omega, x, y)\,\mathrm{e}^{\mathrm{i}(kz-\omega t)} + \mathrm{cc}. \tag{3.2}$$

In Eq. (3.2), cc stands for the complex conjugate of the foregoing expression, and the mode normalization is chosen such that the time-averaged power transported by the electromagnetic wave in Eq. (3.2) is equal to 1 mW when the dimensionless coefficient $A(\omega)$ is equal to one.

In the slowly-varying amplitude approximation it can be shown [19, 30] that the idler and signal amplitudes satisfy the following coupled differential equations (propagation loss as well as pump depletion are neglected, $A(\omega_p, z) = \text{const}$; spatial dependency omitted)

$$\frac{\partial A(\omega_i)}{\partial z} = \mathrm{i}\gamma_i A(\omega_p) A^*(\omega_s) \tag{3.3}$$

$$\frac{\partial A(\omega_s)}{\partial z} = \mathrm{i}\gamma_s A(\omega_p) A^*(\omega_i) \tag{3.4}$$

where we find for the second-order field interaction factor (using Einstein's summation convention and dropping the spatial coordinates for simplicity)

$$\gamma_i = \frac{\epsilon_0 \omega_i}{\int \mathrm{d}x\mathrm{d}y \left[\mathcal{E}^+(\omega_i) \times \mathcal{H}^+(\omega_i)\right]_z} \int \mathrm{d}x\mathrm{d}y\, \chi_{lmn}^{(2)}\, \mathcal{E}_l^-(\omega_i) \left[\mathcal{E}_m^+(\omega_s)\right]^* \mathcal{E}_n^+(\omega_p) \tag{3.5}$$

$$\gamma_s = \frac{\epsilon_0 \omega_s}{\int \mathrm{d}x\mathrm{d}y \left[\mathcal{E}^+(\omega_s) \times \mathcal{H}^+(\omega_s) \right]_z} \int \mathrm{d}x\mathrm{d}y\, \chi^{(2)}_{lmn}\, \mathcal{E}^-_i(\omega_s) \left[\mathcal{E}^+_m(\omega_i) \right]^* \mathcal{E}^+_n(\omega_p) \tag{3.6}$$

where $\mathcal{E}_i^{+(-)}$ represents a forward (back) propagating mode. Equation (3.5) is identical to Eq. (1.37) if one replaces the indices $(1,2,3) \rightarrow (s,p,i)$.

The solution of the coupled equations (3.5) and (3.6) for the case of zero idler at $z = 0$, $A(\omega_i, z = 0) = 0$, is given by

$$A(\omega_s, z) = A(\omega_s, z=0)\cosh(\kappa z) \rightarrow \tfrac{1}{2} A(\omega_s, z=0) e^{\kappa z} \text{ for } z \rightarrow \infty \tag{3.7}$$

$$A(\omega_i, z) = \mathrm{i}\sqrt{\frac{\gamma_i}{\gamma_s}} \frac{A(\omega_p)}{|A(\omega_p)|} A^*(\omega_s, z=0)\sinh(\kappa z) \tag{3.8}$$

where

$$\kappa = |A(\omega_p)| (\gamma_s \gamma_i)^{1/2} \tag{3.9}$$

determines the optical (amplitude) gain. A convenient quantity [75] for representing the device performance is the normalized conversion efficiency, which is defined by

$$\eta = \lim_{z \to 0} P_i(z) / \left(P_s(0) P_p(0) z^2 \right) = \gamma_s \gamma_i \text{ /mW} \tag{3.10}$$

where $P_{i,s,p}(z)$ is the power of the different lightwaves at position z.

For the sake of illustration, we now estimate the nonlinear susceptibility for the organic material M1 which we already used in SOH modulators [3]. This material consists of chromophores dispersed in a polymer matrix, is commercially available at GigOptix [3, 50] and shows an electro-optic coefficient of $r_{33} = 70$ pm/V at the wavelength of 1550 nm. Nonlinear polymers can efficiently be poled in silicon slot waveguides [3] and have reached very high stability, as it has been recently certified by Telcordia [69]. Unfortunately however, for the frequencies involved, no data are available for the nonlinear susceptibility $\chi_{lmn}(\omega_i; \omega_s, \omega_p)$ of the material M1. We adopt therefore the approximation $|\chi_{lmn}| = \delta_{1l}\delta_{1m}\delta_{1n} n^4 |r_{lmn}|/2$ [90], which leads for $n = 1.6$ to $|\chi_{111}| = 230$ pm/V. Assuming further that the nonlinear susceptibility is non-zero only inside the slot, we evaluate numerically the integrals Eq. (3.5), (3.6) for the geometry considered in Fig. 17, and find

$$\gamma_s \approx \gamma_i \approx 16.9 \text{ m}^{-1} \tag{3.11}$$

which correspond to an impressive normalized conversion efficiency

$$\eta = 29 \text{ W}^{-1}\text{cm}^{-2} \quad \left(2900\%\ \text{W}^{-1}\text{cm}^{-2} \right) \tag{3.12}$$

As an example, assuming a CW pump power of 20 dBm, i.e. $A(\omega_p) = 10$, Eq. (3.7) and (3.11) lead to $\kappa = 1.69$ cm^{-1}, which corresponds to a power gain equal to 14.7 dB/cm in the limit of long device length. As a second example, assuming 20 dBm CW input pump power, −10 dBm signal input power, no idler at the input and neglecting losses, Eq. (3.8) implies that after propagating through a 1 cm long waveguide the idler has a power of 0.68 mW (−1.7 dBm), and the signal has a power of 0.78 mW (−1.1 dBm).

We observe that the assumed nonlinear susceptibility of 230 pm/V is a very conservative value. In fact, nonlinear susceptibilities of 354 pm/V have already been measured at optical frequencies in nonlinear polymers [95], and this value increases to 580 pm/V for organic crystals [90]. Moreover, $\chi^{(2)}$ values up to 830 pm/V have already been considered in the context of SOH waveguides [39], and values higher than 3000 pm/V are envisaged in future polymer systems [21]. By using a material with ten times larger nonlinearity, an unprecedented high efficiency of $\eta = 290\,000\ \%\ \mathrm{W^{-1}cm^{-2}}$ could be obtained, or equivalently, 100 times smaller pump powers would lead to the same optical gain. Moreover, the damage threshold of single slot SOH waveguides having much smaller cross-sections is larger than 16 dBm for CW operation [55], suggesting that a pump power of 20 dBm will be below the damage threshold of the device.

It is worth noticing that two-photon absorption (TPA) does not limit the performance of the device. Indeed, even assuming that the entire optical field was concentrated in the 0.4 $\mu\mathrm{m}^2$ silicon cross-section, 20 dBm of pump power correspond to an intensity $I = 25\ \mathrm{MW/cm^2}$. This value, combined with a TPA coefficient $\beta_{\mathrm{TPA}} = 1$ cm/GW [95], corresponds to an absorption coefficient as low as $\beta_{\mathrm{TPA}}I = 0.025\ \mathrm{cm^{-1}}$ (0.1 dB/cm). Free-carrier absorption (FCA) does not constitute a problem either, since it settles in at even higher powers than TPA [47]. Similarly, also third-order nonlinear effects due to the Kerr nonlinearity of silicon can be neglected, since TPA would otherwise be significant as well [96].

We further observe that the silicon dioxide substrate has a (bulk) propagation loss smaller than 2 dB/cm for wavelengths up to 3.6 μm [76, 97] and the silicon itself is transparent in an even larger spectral domain [76]. Also the roughness of the silicon waveguide might induce scattering losses [98], but values below 7 dB/cm have already been demonstrated in (even smaller) single and multiple slot waveguides [99, 100]. The different propagation loss mechanisms will decrease the performance of the device, but because of a gain of 14 dB/cm at 20 dBm pump power, this will not lead to any fundamental change in our discussion.

3.6 Conclusion

In the present work we propose a silicon waveguide concept suited for three-wave mixing. For the first time it is shown that the necessary phase-matching is possible in a silicon-organic hybrid (SOH) waveguide. This could be achieved by dispersion engineering. Our device has high conversion efficiency, is based on standard materials and technologies, and allows all-optical signal processing, mid-IR generation, and lowest-noise optical parametric amplification with small optical pump powers.

3.7 Appendix

A. Mode conversion.

The modes involved in the nonlinear process described above can be excited by taking advantage of a mode converter as the one shown in Fig. 19, which was developed for this purpose. This mode converter acts differently for different wavelengths. At the wavelength of 1550 nm the fundamental (quasi-)TE-mode of the slotted waveguide is converted to the TE_{40}-mode of the double slot waveguide (power transmission coefficient $|S_{21}|^2 = -2$ dB), while at the wavelength of 3100 nm the fundamental mode TE_{00} of the double slot waveguide is excited with $|S_{21}|^2 = -0.7$ dB. This mode converter can be used at the input as well as at the output of the double slot waveguide in order to operate only with the fundamental mode in all the remaining parts of the photonic circuit. The minimum feature size is 100 nm (size of waveguide tip), meaning that e-beam fabrication is not required, and standard 193 nm DUV technology is sufficient.

There are many other mode converter schemes that may be used in order to convert the pump and signal into the respective higher-order modes. Mode converters can for instance be built using the multimode interference (MMI) coupler from Ref. [101], specially designed Bragg gratings, or holograms.

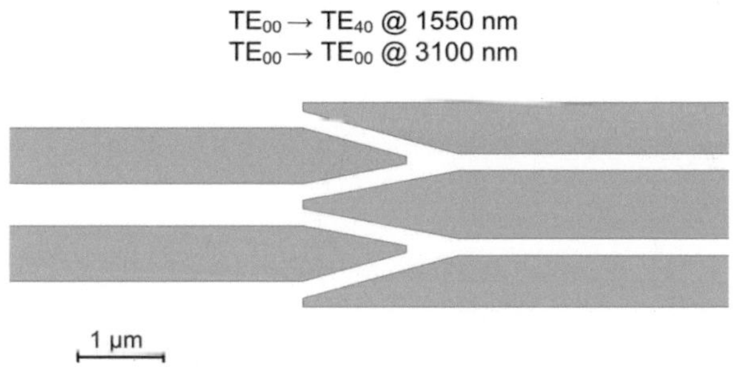

Fig. 19. Slot to double slot mode converter. Light at wavelengths 1550 nm and 3100 nm is coupled in the slotted waveguide on the left. At the wavelength 1550 the fundamental mode of the slotted waveguide is converted into the TE_{40} mode of the double slot waveguide (S_{21} = -2 dB), while at the wavelength of 3100 nm the fundamental mode TE_{00} of the double slot waveguide is excited (S_{21} = 0.7 dB).

B. Third-order nonlinearity vs. second-order nonlinearity

For comparing third-order with second-order nonlinearity we calculate the electric field strength required for creating a certain polarization. The optical response of a material can be described by expanding the polarization $P(t)$ as a power series of the electric field strength $E(t)$. For simplicity we represent the vector fields P and E by scalar quantities,

$$P(t) = P^{(1)}(t) + P^{(2)}(t) + P^{(3)}(t) + \cdots, \qquad P^{(q)} = \chi^{(q)} E^q \tag{3.13}$$

The second-order polarization $P^{(2)}$ is always larger than the third-order polarization $P^{(3)}$ if the electric field is smaller than the critical field

$$E_c = \frac{\chi^{(2)}}{\chi^{(3)}} \tag{3.14}$$

If we now substitute $\chi^{(2)} = 230$ pm/V and $\chi^{(3)} = 10^5$ pm^2/V^2 (this value corresponds to the third-order nonlinear organic molecule DDMEBT [55, 102] which has previously been used for frequency conversion in SOI slot waveguides [55] and has one of the highest Kerr nonlinearities at 1.5 µm [103]), we find that the critical electric field is $E_c = 2.3\ 10^9$ V/m.

For the hypothetical case of a plane wave in vacuum, this field corresponds to an (enormous) intensity of $I = \varepsilon_0 c\,|E|^2 = 1.4\ 10^{16}$ W/m^2, or 140 W on an area of 100×100 nm^2. Since practical devices operate at intensities significantly smaller than the latter [77], $\chi^{(2)}$ waveguides will be more efficient than their $\chi^{(3)}$ counterparts.

4 Silicon-Organic Hybrid (SOH) Phase-Shifters with Liquid Crystal Cladding

4.1 Abstract

The work presented is this chapter extends the SOH modulator technology to other materials. The combination of a silicon slot waveguide with a liquid crystal is explored here for the first time both theoretically and experimentally. The result is a phase-shifter with $V_\pi L = 0.085$ Vmm and nW-power consumption – the smallest values ever reported for silicon photonics. For a device length of 1.7 mm, a drive voltage of 5 V leads to a 35 π phase shift.

This chapter is based on the following publications:

Liquid Crystal Phase Shifter on the SOH Platform with Ultra-Low Power Consumption

Luca Alloatti, Joerg Pfeifle, Jorge Mendez, Wolfgang Freude, Juerg Leuthold, Christian Koos

Optical Fiber Communication Conference (OFC'12), Paper OTu1I.5, Los Angeles (CA), USA, March 2012

The same experimental data is published also in Optics Express, Vol. 20, Issue 14, pp. 15359-15376, July 2012

4.2 Introduction

Phase shifters are essential for a large variety of applications in silicon photonics such as optical matrix switches, polarization controllers [104], tunable filters [105, 106], and delay-interferometers, e. g., in fast Fourier transformers [107]. To enable dense integration, phase shifters must be compact, power-efficient, and tuning voltages should be below 1 V.

Thermal phase shifters exploit the high thermo-optic coefficient of silicon by heating silicon nanowires with a constant current flow. These devices enable phase shifts of few π [104, 105], however, their viability is limited by their power dissipation and the associated limitations in integration density.

Liquid crystals offer the potential of maintaining a phase shift with negligible power consumption [108-111], but at present, low phase shifting efficiencies lead to long devices or high driving voltages.

In this paper we demonstrate a phase-shifter with a record voltage-length product of $V_\pi L = 0.085$ Vmm – a 50-fold improvement compared to the present state of the art. This is obtained by combining liquid crystals with slot-waveguides within the silicon-organic hybrid platform [46].

4.3 Design and Fabrication

Liquid-crystals are highly birefringent materials, which can reorient under the influence of an electric field. This makes them ideal candidates for electro-optic applications. Strip-loaded slot waveguides enable efficient interaction of the guided light with a functional cladding, and at the same time allow generating high modulating electric field strengths [3]. The combination of these advantages is expected to lead to highly-efficient phase shifters with purely capacitive input impedances and therefore ultra-low power consumption.

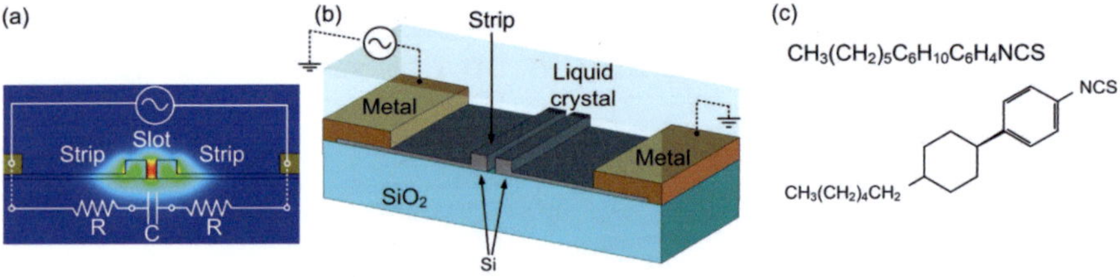

Fig. 20. Strip-loaded slot waveguide. (a) Cross-section of the optical waveguide. The optical active region is connected to the metal electrodes by thin silicon strips (strip load). The color code represents the electric field magnitude of the optical mode; the light is concentrated inside the slot. Also indicated is the lumped equivalent circuit of the device. The slot is represented by a capacitance and the strip-load by a resistance. (b) Sketch of the waveguide. The liquid crystal is assumed to cover the waveguide homogeneously, and to entirely fill the slot. The nominal rail and slot widths of the waveguide are 120 nm and 240 nm, respectively. (c) The chemical and molecular structure of our nematic liquid crystal.

Fig. 20 depicts the strip-loaded slot waveguide used in the experiment. The external voltage is applied to the metallic electrodes located at a sufficient distance from the optical slot waveguide. The applied voltage drops entirely across the 120 nm wide slot, creating a strong electric field which can reorient the director of the liquid crystal deposited in the slot. The resistive silicon strips together with the capacitive slot act as an *RC*-circuit, whose time constant is negligible because of the slight doping of the silicon strips [3]. The phase shifter region is 1.7 mm long and has been fabricated entirely in a silicon fab, see Ref. [3] for a detailed description.

The liquid crystal we chose is the nematic crystal 1-(trans-4-Hexylcyclohexyl)-4-isothiocyanatobenzene distributed by Sigma-Aldrich (linear formula $CH_3(CH_2)_5$ $C_6H_{10}C_6H_4NCS$, product number 366854). This material is liquid at room temperature, and its molecules are asymmetric as can be seen from Fig. 20(c). A small amount of it was drop-casted on the waveguide in normal atmosphere. Because of their asymmetry the molecules possess a permanent dipole moment which is responsible of the mechanical reorientation under the influence of an external electric field.

4.4 Device Characterization

Since the DUT consists of a single phase shifter, we mounted the device inside one arm of a fiber-based Mach-Zehnder interferometer as sketched in Fig. 21. The interference pattern is recorded on an oscilloscope and the phase shift was derived from it. The optical insertion loss is measured with a photodetector placed after a 3 dB directional coupler at the output of the device. The voltage of a function generator is applied to the chips by micro-needles. The measurement results are depicted in Fig. 22.

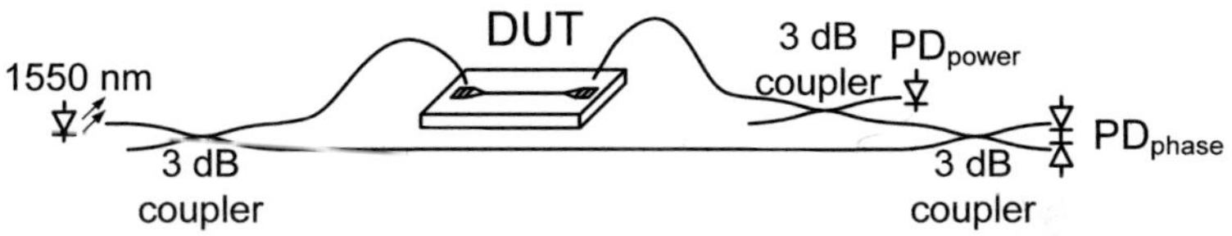

Fig. 21. Measurement setup. Light from a 1550 nm laser source is split by a 3 dB directional coupler. One part is launched in the DUT by grating couplers. At the output of the chip the optical power is monitored by an additional 3 dB coupler and a photodiode (PD_{power}). Light coming from the DUT interferes with light from the second arm of the Mach-Zehnder interferometer. The interference pattern is recorded with a balanced photodetector and reveals the effective phase shift.

4.5 Results

The phase shift generated by the liquid crystal is a nonlinear function of the applied voltage. In Fig. 22(a) we report the phase shift as a function of time when a triangular signal having 5 V amplitude is applied to the chip. When the voltage crosses the zero line, the phase shift reaches a minimum. The phase is symmetric with respect to positive and negative voltages, consistently with the geometrical symmetry of the device. The maximum phase shift obtained in this case is about $35\,\pi \approx 101$ rad, and the maximum differential phase shift equals $20\,\pi$ rad/V. For the 1.7 mm long device this corresponds to a voltage-length product of $V_\pi L = 0.085$ Vmm. This value is by far the lowest recorded on a silicon chip with liquid crystal cladding, polymer cladding, or for a pn-junction [3, 37].

If a DC bias voltage is superimposed to the AC signal, the phase responds approximately linearly to the applied voltage; this is shown in Fig. 22(b) for a DC potential of 8 V and an AC amplitude of 1 V. Additionally, we observed a voltage-dependent insertion loss (blue curve in Fig. 22), which is attributed to the light scattering occurring at the liquid crystal domain boundaries [112, 113].

The resistance of the device depends on the applied voltage, and varies between 1.7 GΩ for 1V bias, 280 MΩ for 5 V bias, and 3 MΩ for 10 V bias. The ground and signal metallic electrodes of the device have an area of 0.12 mm^2. An even higher resistance is expected when using smaller electrodes, and shorter devices. By operating the device at 5 V DC the electrical power consumption of our device is 90 nW. This value is about six orders of magnitude smaller than the power consumption of thermal silicon phase shifters.

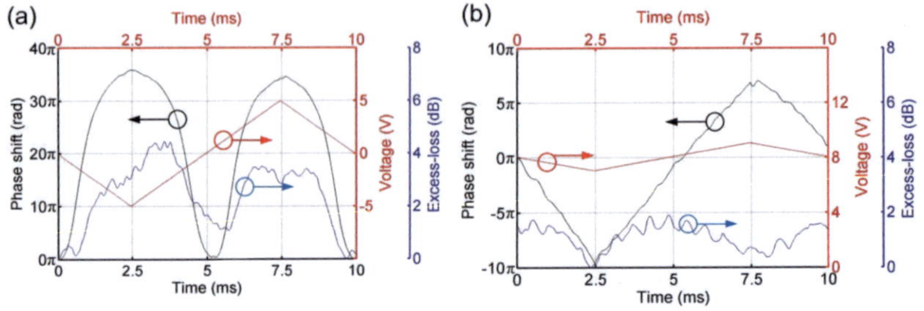

Fig. 22. Phase response of the DUT. (a) When applying a 100 Hz triangular signal with an amplitude of 5 V (red curve), the phase shift responds nonlinearly to the voltage (black curve). The maximum phase shift achieved in this case is $35\,\pi \approx 101$ rad. The maximum differential phase shift equals $20\,\pi$ rad/V, which is equivalent for the 1.7 mm long device to a voltage-length product $V_\pi L = 0.085$ Vmm. (b) Response of the DUT when a 1 V amplitude triangular signal is added to an 8 V DC bias. The differential phase shift in this case is $8\,\pi$ rad/V. The blue curves represent the variation of the optical loss for different liquid-crystal orientations. The rapidly-varying oscillations superimposed to the loss signal are attributed to Fabry-Perot resonances due to input and output reflections of the waveguide, when the optical length is varied by the modulation. Additional noise is attributed to the mechanical vibrations of the measurement setup.

Because of the high phase shifting efficiency, a device length of 100 μm is sufficient for most applications. In this case a π-shift could be obtained with a voltage swing smaller than 1 V and the voltage-dependent loss would be smaller than 0.3 dB. The typical propagation loss in a 100 μm long slot-waveguide is 0.15 dB [55]; and by adding the liquid crystal this value rises by 1.6 dB. We expect that a systematic study of the liquid crystals available on the market will lead to significant smaller propagation losses [112].

4.6 Conclusions

We showed for the first time that liquid crystals can be combined with a silicon slot waveguide. We demonstrated a record voltage-length product of $V_\pi L = 0.085$ Vmm, and we proved that phase shifters with a power consumption in the nW range are feasible.

4.7 AC Driving Signal

This section is based on the following publication:

Silicon-organic hybrid devices

L. Alloatti, D. Korn, J. Pfeifle, R. Palmer, S. Koeber, M. Baier, R. Schmogrow, S. Diebold, P. Pahl, T. Zwick, H. Yu, W. Bogaerts, R. Baets, M. Fournier, J. M. Fedeli, R. Dinu, C. Koos, W. Freude, J. Leuthold

SPIE Photonics West (OPTO-SPIE'13), San Francisco (CA), USA, paper 8629-24; Feb. 2-7, 2013, **[invited]**

As we have seen in Section 4.4, in our first experiments the liquid crystal was driven with a DC field and suffered from ion drift. We recently solved this problem by using the commercially available liquid crystal E7 and the same waveguide as described in Section 4.3. By modulating the envelope of a 100 kHz square wave, the voltage-length product could be decreased by an additional factor two, Fig. 23. The alternating voltage prevents ion migration. This technology is commonly used in LC displays.

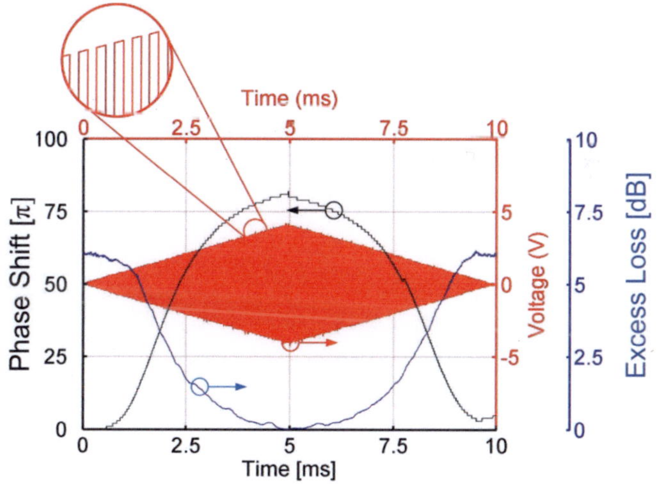

Fig. 23. Liquid crystal phase shifter. Phase-shift, applied voltage and excess optical loss vs. time. The envelope of a 100 kHz square wave is modulated with a triangular function having frequency 100 Hz. The maximum electrical amplitude is 4V. The maximum phase shift achieved is about 80 π. The device length is 1.7 mm.

5 Optical Absorption in Silicon due to Charge Inversion/Accumulation

L. Alloatti, M. Lauermann, C. Sürgers, C. Koos, W. Freude, J. Leuthold

Accepted for publication in APL

5.1 Abstract

We determine the optical losses in gate-induced charge accumulation/inversion layers at a Si/SiO_2 interface. Comparison between gate-induced charge layers and ion-implanted thin silicon films having an identical sheet resistance show that optical losses can be significantly lower for gate-induced layers. For a given sheet resistance, holes produce higher optical loss than electrons. Measurements have been performed at $\lambda = 1550$ nm.

5.2 Introduction

Small optical losses are essential for silicon photonics devices. However, for keeping *RC* time constants low and bandwidths high [4, 24, 27], an increasing number of components, like high-speed electro-optic modulators [18, 68] or photodetectors, require highly conductive electrodes, which in turn create high free-carrier absorption (FCA).

Even though there are several methods to increase the conductivity ($1 / \rho$) of a semiconductor, so far only doping has been considered in silicon photonics. For thin conductive films the sheet resistance $R_s = \rho / d$ is the relevant number when the film thickness d is given; the resistance of a sheet of length L and width w then is $R = R_s L / w$. However, while the optical loss via free-carrier absorption increases linearly with the doping level [59], the conductivity grows less than linearly [114, 115], mainly because of electron-impurity scattering [58]. This confronts designers with a trade-off problem between device speed and insertion loss, properties which both are of primary technological importance [18, 26, 116]. A method for improving the conductivity of silicon structures without an undue increase of optical loss would therefore be highly desirable.

A possible solution for minimizing optical losses while maintaining a high electrical conductivity has recently been suggested [3]. In this publication the bandwidth of a silicon-based modulator has been dramatically increased by exploiting an electron accumulation layer induced by a gate voltage. The optical loss created by such an accumulation layer was very small. To the best of our knowledge however, the optical loss caused by inversion/accumulation layers in silicon has never been measured systematically.

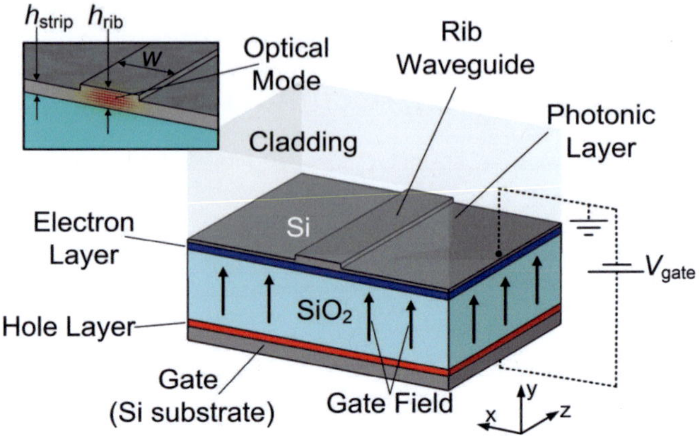

Fig. 24. Cross-section of the silicon chip used for optical loss vs. gate field measurements. The waveguide is located on top of a low-refractive index substrate (SiO_2). The optical quasi-TE_{00} mode is mainly confined in the upper lowly-doped (ideally intrinsic) silicon layer, and extends slightly into the silicon-dioxide substrate and into the 0.8 µm thick PMMA cladding. A gate voltage (V_{gate}) is applied between the silicon substrate and the grounded photonic layer. In the example shown in the picture, the gate voltage is positive, so that a hole charge layer (marked in red) is formed at the interface between gate and oxide, while an electron layer (marked in blue) is formed between the waveguide and the oxide. This electron layer reduces the sheet resistance in the photonic silicon layer, but also increases the optical losses. The rib height is $h_{rib} = 220$ nm, the strip height is $h_{strip} = 150$ nm, the rib width is $w = 700$ nm, and the silicon-oxide thickness is $d_{SiO_2} = 2\,\mu$m.

In this work we measure for the first time the optical absorption caused by electron and hole charge accumulation layers at a silicon/silicon dioxide interface. For equal charge carrier concentrations as a result from either impurity doping, or from free carriers e.g. accumulated under the influence of an applied electric field, the optical losses are comparable in both cases. However, due to impurity scattering, the electrical conductivity is smaller for doped layers. Thus, for the same conductivities, the concentration of impurity-generated charges must be higher leading to a higher optical loss. By comparing our measurements with published data for doped layers, we show that the optical power attenuation coefficient for field-induced free carriers and given sheet resistance can be up to 3.5 times smaller.

The chapter is organized as follows: First, we determine the optical loss in a rib waveguide when a gate voltage is applied. Second, we derive a model for describing such losses. Finally, we compare these losses to published data for ion implantation.

5.3 Results

For determining the optical loss of conductive sheets as a function of gate fields, we record the optical transmission of rib waveguides structured on a silicon-on-insulator (SOI) wafer from Soitec, Fig. 24. We measure the transmission as a function of the gate voltage V_{gate} applied to the silicon substrate (the photonic layer is electrically grounded). The fundamental quasi-TE_{00} mode of the waveguide is strongly confined to the photonic layer, Fig. 24. The mode couples to external fibers by standard gratings (see supplemental information for more details). If the gate field across the silicon dioxide $E_{gate} = V_{gate}/d_{SiO2}$ is positive (negative), an electron (hole) layer is formed at the Si/SiO_2 boundary inside the photonic layer, Fig. 24. These charges then lower the sheet resistance of the photonic layer but will also increase the optical losses. Subsequently, we will measure and discuss these losses.

The optical transmission losses have been measured for different waveguide lengths (1, 3 and 10 mm) and the excess optical loss for free-carrier absorption has been extracted, black curve in Fig. 25. As expected, the optical loss increases for large positive or large negative gate fields. The minimum optical loss occurs at $E_{gate} = -0.018$ V/nm instead at $E_{gate} = 0$ or at a slightly positive gate field, as one would expect since silicon has a residual *p*-type doping, see supplemental information. This could be attributed to the presence of the PMMA cladding, but also residual impurities could play a role.

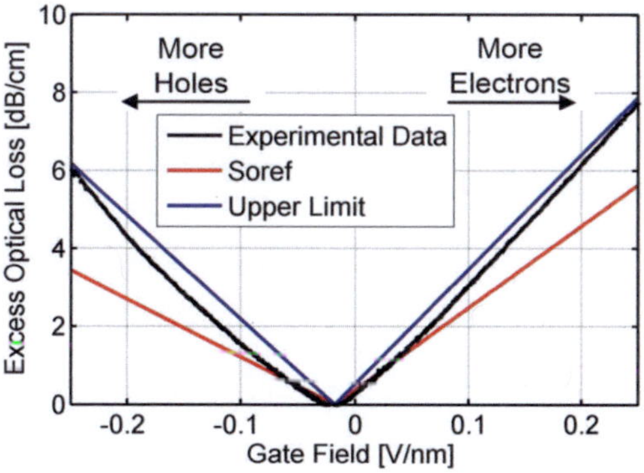

Fig. 25. Measured excess optical loss of a rib waveguide as a function of the applied gate field (black line). For large positive (negative) gate fields, an electron (hole) layer forms in the photonic layer, and the optical loss increases. The red curve shows the calculated losses according to Eqs. (5.1),(5.2) and (5.3), which were empirically derived by Soref [47, 59] from doping measurements. The blue curve represents our measurements in a linear worst-case approximation.

We now compare the measured loss data in Fig. 25 as induced by applying a gate field with losses predicted due to free carrier absorption with doping. For this we follow Soref [47, 59]. According to Soref the optical power attenuation coefficient at position $\vec{r}$ due to free carriers increases linearly with the concentration N_e and N_h of electrons and holes respectively,

$$\alpha(\vec{r}) = C_e N_e(\vec{r}) + C_h N_h(\vec{r}). \tag{5.1}$$

In the case that the free carriers stem from impurity ionization only, the loss constants C_e and C_h in Eq. (5.1) have been empirically determined [47, 59] at the wavelength $\lambda = 1550$ nm,

$$C_e^{(\text{ion})} = 8.5\times10^{-22}\,\text{m}^2, \tag{5.2}$$

$$C_h^{(\text{ion})} = 6.0\times10^{-22}\,\text{m}^2. \tag{5.3}$$

In contrast to Soref, here the free-carrier concentration in the lowly-doped (ideally intrinsic) photonic layer depends on the gate field. If charges are injected in intrinsic silicon by applying a field, different constants might apply. Because these injected carriers do not experience impurity scattering [58], optical losses could be smaller. On the other hand, when charges are injected by a gate voltage, charge densities close to 10^{20} cm^{-3} can appear in the vicinity of the Si/SiO_2 interface – a concentration regime where electron-electron scattering cannot be neglected [58]. Additionally, also the roughness of the Si/SiO_2 interface might play a role [117]. Therefore it is not clear *a priori* whether Soref's constants overestimate or underestimate the actual loss.

For calculating the waveguide attenuation coefficient α_{wg} of the quasi-TE_{00} mode using Eq. (5.1), we first define in a Cartesian coordinate (x, y, z)- system (Fig. 24) the electric field components in the corresponding directions denoted by $n = (1, 2, 3)$, and separate a dimensionless complex amplitude $A(z)$ from the modal field $\mathcal{E}_n(x, y)$. A monochromatic wave propagating along the z-direction with propagation constant k and an angular frequency ω is described by

$$E_n(x,y,z,t) = A(z)\mathcal{E}_n(x,y)e^{i(kz-\omega t)} + \text{cc}, \tag{5.4}$$

$$H_n(x,y,z,t) = A(z)\mathcal{H}_n(x,y)e^{i(kz-\omega t)} + \text{cc}. \tag{5.5}$$

In the slowly-varying amplitude approximation it can be shown [30] that the waveguide attenuation coefficient $\alpha_{\text{wg}} = -\left(1/|A(z)|^2\right)\partial |A(z)|^2/\partial z$ is given by (vacuum speed of light c, vacuum permittivity ϵ_0)

$$\alpha_{\text{wg}} = \frac{\epsilon_0 c}{\int \mathrm{d}x\mathrm{d}y\left(\mathcal{E}_1\mathcal{H}_2 - \mathcal{E}_2\mathcal{H}_1\right)} \int \mathrm{d}x\mathrm{d}y\, n(x,y)\alpha(x,y)\left(\mathcal{E}_1^2 + \mathcal{E}_2^2 - \mathcal{E}_3^2\right), \tag{5.6}$$

where $n(x,y)$ and $\alpha(x,y)$ are the local refractive index and the local attenuation coefficient Eq. (5.1), respectively. We notice that the last negative sign in Eq. (5.6) is due to the fact that if $\mathcal{E}_{1,2}(x,y)$ is chosen to be real, the quantity $\mathcal{E}_3(x,y)$ will be imaginary.

Next, we substitute Eq. (5.1) in Eq. (5.6). Since the charge distribution varies only along y, the local power attenuation coefficient $\alpha(x,y) = \alpha(y)$ does not depend on x (see coordinate system in Fig. 24). For the gate voltages considered in this work, our simulations show (see supplemental information) that the electron and hole densities do not exceed 10^{20} cm^{-3}. This means that the variation of the silicon refractive index is less than 0.3 % [59]; we therefore assume that the refractive index is constant, $n(x,y) = 3.48$. If we further assume that $\alpha(y)$ is large only in a thin silicon layer in the vicinity of the Si/SiO_2 interface at $y = 0$, the electric field $\mathcal{E}_n(x,y)$ can be considered constant with respect to y, and the integral in (5.6) becomes

$$\alpha_{\text{wg}} = C_{e(h)}\sigma_{e(h)}\tilde{\Gamma}, \tag{5.7}$$

where $C_{e(h)}$ represents the constant C_e for positive gate voltages or the constant C_h for negative voltages, and $\sigma_{e(h)} = \sigma_{e(h)}(V_{\text{gate}})$ is the corresponding two-dimensional voltage-dependent carrier density (unit m^{-2})

$$\sigma_{e(h)} = \int \mathrm{d}y\, N_{e(h)}(y). \tag{5.8}$$

If the silicon is intrinsic and the charge layer thickness is much smaller than the oxide thickness (as in our case), the two-dimensional carrier densities for electrons and holes ($\sigma_e = \sigma$ for $V_{\text{gate}} > 0$, $\sigma_h = \sigma$ for $V_{\text{gate}} < 0$) can be approximated by

$$\sigma = \epsilon_0 \epsilon_{\text{SiO}_2} \left|V_{\text{gate}}\right| / \left(q d_{\text{SiO}_2}\right), \tag{5.9}$$

where $\epsilon_{\text{SiO}_2} = 3.8$ is the static relative permittivity of silicon dioxide, and q is the elementary charge [33].

The so-called optical interaction factor $\tilde{\Gamma}$ of the waveguide (unit m^{-1}) is defined as

$$\tilde{\Gamma} = \epsilon_0 c n_{\text{Si}} \frac{\int \mathrm{d}x \left(\mathcal{E}_1^2 + \mathcal{E}_2^2 - \mathcal{E}_3^2\right)\big|_{y=0^+}}{\int \mathrm{d}x\mathrm{d}y \left(\mathcal{E}_1 \mathcal{H}_2 - \mathcal{E}_2 \mathcal{H}_1\right)}. \tag{5.10}$$

For evaluating the interaction factor, we perform a numerical integration of the mode field. To this end, the optical mode is computed with commercially available software [28], and we find $\tilde{\Gamma} = 2.63 \mu\text{m}^{-1}$.

Using the constants $C_{e(h)}^{(\text{ion})}$ determined by Soref, Eqs. (5.2) and (5.3), we obtain the red curve in Fig. 25, which was shifted by the experimentally observed offset –0.018 V/nm as discussed before. The constants as used by Soref therefore underestimate the actual optical loss in the inversion/accumulation layer. If Soref's model should be applied nonetheless, its constants must be adjusted. In Fig. 25 the blue straight lines are a better approximation to the measurements and represent an upper loss limit. The corresponding loss constants are

$$C_e^{(\text{upper})} = 1.4\, C_e^{(\text{ion})}, \tag{5.11}$$

$$C_h^{(\text{upper})} = 1.8\, C_h^{(\text{ion})}. \tag{5.12}$$

We therefore conclude that optical loss due to an inversion/accumulation layer can be empirically described by Eq. (5.1) only if the coefficients $C_{e(h)}$ become field-dependent. For gate fields in the range from –0.25 V/nm to 0.25 V/nm, we find that the coefficients $C_{e(h)}$ lie between the values determined by Soref, Eqs. (5.2) and (5.3), and those given in Eqs. (5.11) and (5.12).

The assumption made in deriving Eqs. (5.7) to (5.9), namely that the charge layer is infinitesimally thin, is true only approximately. However, numerical simulations based on industry's standard software ATLAS show that more than 50% of the total charge in intrinsic silicon is located within the first 5 nm from the interface (gate fields larger than 0.1 V/nm or smaller than –0.1 V/nm). If in this range of gate fields we replace the simulated charge distribution by a delta-shaped distribution as is assumed in Eq. (5.9), the calculated waveguide attenuation coefficient would differ by less than 5 %, so we adopt the simpler approach Eqs. (5.7) to (5.9).

Now that we have measured the losses for a given gate field, we are also interested in the sheet conductance for a given gate field. Although several models exist for calculating the sheet resistance of accumulation/inversion layers in silicon [117, 118], we feel the need to measure this quantity for a sample belonging to the same SOI wafer as used for the optical loss measurements. To this end, a 2 mm × 2 mm square has been etched into a 220 nm thick photonic layer. The chip has been placed in a high-vacuum chamber for avoiding contamination of the surface, and to reduce the risk of electrical breakdown (see supplemental information for more details). The resistance is measured by a standard four-point procedure [119], the result of which is shown in Fig. 26. For the highest gate field, $E_{\text{gate}} = 0.25$ V/nm, we measure a sheet resistance of 2.91 kΩ/sq. This value corresponds, using Eq. (5.9), to an effective electron mobility $\mu_{e,\text{eff}} = 1\,/\,(q\sigma R_s) = 394\ \text{cm}^2/\text{V/s}$, a factor 3.63 smaller than the intrinsic electron mobility $\mu_e = 1430\ \text{cm}^2/\text{V/s}$. For the gate field $E_{\text{gate}} = -0.25$ V/nm we obtain a sheet resistance of 10.8 kΩ/sq. We observe that the sheet resistance in inversion/accumulation layers on thermally oxidized silicon (as in our SOI) is a sensitive function of surface roughness and Coulomb scattering with fixed oxide charges [117]. Effective mobilities larger than the one we found are possible [117], leading to potentially higher sheet conductances and lower optical losses for the same gate fields.

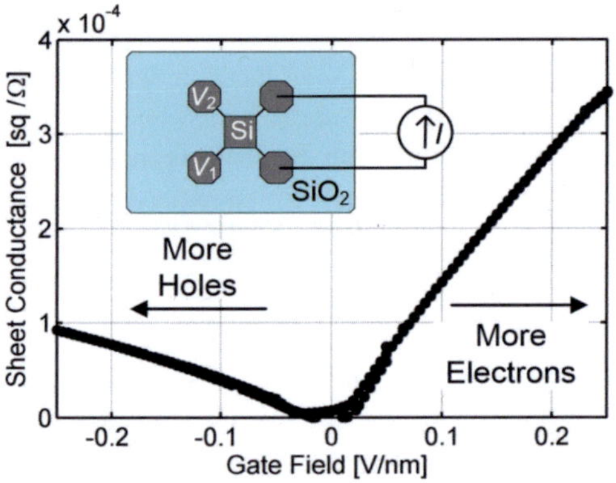

Fig. 26. Sheet conductance (inverse of sheet resistance) vs. gate field. The sample used consists of a 220 nm thick silicon belonging to an SOI wafer identical to the one used for the optical loss measurements. The measurements were performed in vacuum. The 2 mm × 2 mm silicon square as used for the four-point measurement is shown in the inset (the octagons are used for contacting the chip and connect the vertices of the square with 50 μm wide silicon strips). A current I is forced to flow across two adiacent vertices and the voltages V_1 and V_2 are measured on the other vertices. The sheet resistance is calculated as $R_s = 4.5324\,(V_2 - V_1)\,/\,I$ [119].

Finally, we compare optical losses of gate induced sheet layers with losses of ion-implanted layers for a given sheet resistance by using published data on conductivity and optical loss in doped silicon.

In order to get rid of any influence from the waveguide geometry (see interaction factor in Eq. (5.6)), we investigate an optical mode propagating along a doped silicon slab having a thickness d in a lossless background material with the same (real) refractive index $n_{Si} = 3.48$ as silicon. The optical mode is assumed to have dimensions much larger than the slab thickness d, Fig. 27(a).

We start looking at the attenuation coefficient $\alpha_{e(h)}(d, R_s)$ in n-doped (p-doped) silicon slabs of thickness d with a sheet resistance $R_{s,e(h)}$. The sheet resistance $R_{s,e(h)} = (q\mu_{e(h)} N_{e(h)} d)^{-1}$ depends on the electron (hole) mobility $\mu_{e(h)}$, the electron (hole) density $N_{e(h)}$ and the slab thickness d. From tabulated data [120] we extract the required ion concentration (for both n or p impurity type) for a given sheet resistance. We then assume that the ion concentration equals the free-carrier density $N_{e(h)}$ (100 % impurity ionization, see supplemental information). The attenuation coefficient $\alpha_{e(h)}(d, R_s)$ is then given by Eq. (5.7). Since the interaction factor $\tilde{\Gamma}$ depends from the particular optical mode, we set here arbitrarily $\tilde{\Gamma} = 1\mu\text{m}^{-1}$. It is found that the losses depend on the sheet thickness d, Fig. 27(b,c). This is because high impurity concentrations in silicon degrade the carrier mobility. For example, if the slab thickness is decreased, by a factor of ten, the ion concentration must be increased by more than a factor of ten for keeping the same sheet resistance R_s. And in fact, a sheet resistance of, e.g., 3 kΩ/sq can be obtained in a 1000 nm thick slab with an n-type ion concentration of 1.84×10^{16} cm^{-3}, as well as with a concentration of 6.03×10^{17} cm^{-3} (32 times larger) in a 100 nm thick slab, or a concentration of 2.2×10^{19} cm^{-3} (an additional factor 36 larger) in a 10 nm thick slab. For an ion concentration of 2.2×10^{19} cm^{-3}, the electron mobility has decreased by a factor 15 of its intrinsic value $\mu_e = 1430$ cm^2/V/s.

In the limit of large slab thickness, the attenuation coefficient $\alpha_{e(h)}(d, R_s)$ converges to a finite limit which depends on the electron (hole) intrinsic mobility ($\mu_e = 1430$ cm^2/V/s and $\mu_h = 495$ cm^2/V/s) as well as on the loss constants Eqs. (5.2) and (5.3). We denote this limit with $\alpha_{e(h)}(\infty, R_s)$. The ratio $\alpha_h(\infty, R_s) / \alpha_e(\infty, R_s)$ equals

$$\alpha_h(\infty, R_s) / \alpha_e(\infty, R_s) = C_h^{(\text{ion})} / C_e^{(\text{ion})} \times \mu_e / \mu_h = 2.04. \tag{5.13}$$

Since this value is larger than one, we recover the well-known result that for moderate doping levels and a fixed sheet resistance holes produce more optical loss than electrons. Next, we derive the attenuation coefficient $\alpha_{e(h)}(d, R_s)$ in the case of a gate-induced conductive sheet with a sheet resistance R_s. Since accumulation and inversion layers are a few nanometers thick only, the optical losses for layers thicker than, e. g., 5 nm must be independent of the slab thickness. For a given sheet resistance we derive the required gate voltage from the experimental data plotted in Fig. 26. The gate field in turn determines the optical losses according to the data in Fig. 1. The results are plotted in Fig. 27(b,c). We see that optical losses are independent of the slab thickness as stated above. Additionally, according to Fig. 27, holes produce more optical loss than electrons as is the case for ion implantation. But most importantly, one may observe that for thin slabs and for the same sheet resistance smaller optical loss can be obtained by injecting electrons with a gate voltage instead of using ion implantation.

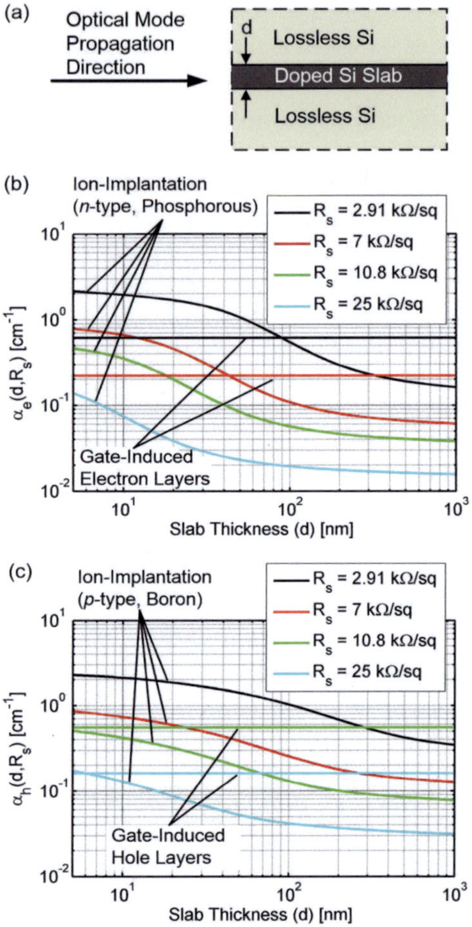

Fig. 27. Optical power attenuation coefficient of a mode (a) propagating along a silicon slab vs. slab thickness for a constant sheet resistance R_s, (b) for *n*-doping and an electron gate-induced layer and (c) for *p*-doping or gate-induced hole layers. The differently coloured lines represent different sheet resistances (legend) of the slab waveguide. If the carrier mobility in silicon would not depend on the doping levels, the curves for ion-implantation would be constant lines. The sheet resistance of gate-induced layers does not depend on the thickness of the silicon slab. For electrons, smaller optical losses can be obtained with gate-induced charge layers, especially for small thicknesses.

In conclusion, we have measured for the first time optical loss of inversion/accumulation layers in silicon, and loss coefficients have been determined. We find that the optical losses of thin sheet layers are smaller when the resistance is reduced by an inversion/accumulation layers rather than by doping silicon slabs.

We acknowledge support by the DFG Center for Functional Nanostructures (CFN), the Helmholtz International Research School of Teratronics, the Karlsruhe School of Optics and Photonics (KSOP), the EU-FP7 projects SOFI (grant 248609), the BMBF joint project MISTRAL and the European Research Council (ERC Starting Grant 'EnTeraPIC', number 280145). We are grateful for technological support by the Karlsruhe Nano-Micro Facility (KNMF). We further acknowledge the Alfried Krupp von Bohlen und Halbach Foundation, the Deutsche Forschungsgemeinschaft and Open Access Publishing Fund of Karlsruhe Institute of Technology.

5.4 Supplemental information

A. Waveguide fabrication and passivation.

The substrate used is a 10 × 14 mm standard photonic SOI chip purchased from SOITEC having a 220 nm thick device layer, and a buried-oxide (BOX) thickness of 2 µm. The silicon is specified as *p*-type with a residual impurity concentration of $N_A \approx 10^{-15}\,\mathrm{cm}^{-3}$ with a (100) crystal orientation. For avoiding electrical breakdown when applying high gate voltages, the 220 nm thick silicon in the first 1 mm from the chip's edges was removed by SF_6 reactive-ion etching (RIE) using a resist mask (ma-N 1410 available at micro resist technology GmbH). After rinsing with acetone, isopropanol and deionized water, the chip was cleaned with oxygen plasma and placed on a hotplate for 5 min at 150 °C for water desorption. The adhesion promoter Ti-prime (available from MicroChemicals GmbH), was spin coated at 4000 rpm for 30 s, acceleration 1000 rpm/s. The adhesion promoter was activated on a hotplate at 120 °C for 2 min. This transforms the organic molecules containing titanium in a non-closed atomic layer of TiO_2. The negative-tone resist ma-N 2401 (available from micro resist technology GmbH) was spin coated at 3000 rpm for 30 s, acceleration 3000 rpm/s, for achieving a nominal resist thickness of 100 nm. The chip was exposed to e-beam lithography on a JEOL system with 50 keV energy, 70 µC/cm^2 dose and 4 nm step size. Waveguides were fabricated along the [110] direction. The resist was subsequently developed for 12 s in Dev-525 (available at MicroChem GmbH) and rinsed in deionized (DI) water. The chip was subsequently immersed for 2 s in 1 % hydrofluoric acid (HF) for removing the thin TiO_2 layer (which is highly resistant against SF_6 etching) and rinsed in DI water. Within few minutes the chip was inserted in the reactive-ion etching chamber for avoiding the formation of native oxide. Reactive-ion etching (RIE) of silicon was made for 7 s in a Sentec Etchlab 200 RIE with SF_6 at a pressure of 0.1 Pa, gas flow 5 sccm, RF power 500 W. This resulted in 70 nm etch depth, based on the etch rate determined immediately before by etching the back-side of an identical SOI chip for two hours, and measuring the thickness difference with a micrometer; an etch rate of 10.0 nm/s was determined. The residual resist was removed with acetone. Subsequently, the sample was carefully cleaned for removing surface contaminations. To this end the chip was immersed for 20 s in 1 % hydrofluoric acid (HF), and rinsed in ultraclean water (part-per-trillion grade from Carl Roth, ROTIPURAN Ultra). The chip was then cleaned for 15 min in standard cleaning 1 (SC1) solution [121] at 75 °C, rinsed twice in ultraclean water, immersed 10 min in standard cleaning 2 (SC2) solution [121] and rinsed again in ultraclean water; all chemical used during SC1 and SC2 (H_2O, NH_4OH, H_2O_2, HCl were of part-per-trillion grade from Carl Roth, ROTIPURAN Ultra). The chip was then placed 10 min on a hotplate at 150°C for water desorption before it was finally spin-coated at 4000 rpm with a 0.8 µm thick layer of resist (PMMA 950 k in anisole, AR-P 672.08 available from Allresist). The resist was baked on a hotplate at 180 C for 5 min and finally tested. The resist cladding protects the optical waveguide from dust during the optical measurement, has a known refractive index ($n = 1.48$), and can be easily scratched with a needle for creating an electrical contact with the silicon photonic layer. The final device is schematically depicted in Fig. 1.

B. Measurement details.

As optical source we used amplified spontaneous emission (ASE) from an erbium doped fibre amplifier (EDFA) filtered with 1 nm bandwidth around the wavelength of 1550 nm. This way any artifacts in the optical transmission deriving from multiple reflections at the grating couplers and waveguides terminations could be avoided. The gate voltage was ramped several times from −500 V to +500 V and vice versa with 1 V steps every 0.2 s, such that any possible hysteresis could be monitored and excluded.

For the sheet resistance measurement the four contacts were silicided with nickel for guaranteeing a good electrical contact with the inversion/accumulation layer across the 220 nm thick silicon. To this end, the sample was covered with a stainless-steel mask having four holes at contact locations, and heated to 600 °C in 10^{-2} mbar Ar atmosphere. A layer of 1 nm of Ni was sputtered, then the sample was cooled to 400 °C, and 400 nm of Ni were sputtered for creating a mono nickel silicide; finally the sample was cooled and measured under vacuum and at room temperature with a Lake Shore 370 AC resistance bridge.

C. Impurity ionization percentage.

In calculating the data for Fig. 27(b,c), 100 % impurity ionization was assumed. This assumption is justified by the fact that Eqs. (5.1), (5.2) and (5.3) were derived from data reporting optical loss for given impurity concentrations, and assuming 100 % impurity ionization [59]. Additionally, it has been shown that the ionization of phosphorous donors is indeed close to 100 % for the entire impurity density range from 10^{15} cm^{-3} to 10^{20} cm^{-3}. The ionization reaches a minimum, but remains still larger than 90 %, close to an impurity concentration of 10^{18} cm^{-3} [122] [123].

6 Summary and Outlook

In this thesis two types of phase shifters have been built and tested experimentally. Both of them established new records in silicon photonics, the first one in terms of speed, and the second one in terms of driving voltage and power consumption. A third type of device has been proposed for achieving three-wave mixing on the silicon-organic hybrid (SOH) platform; this is the first time that such a device has been proposed for the SOH platform.

All these devices can be further developed. In the following an outlook on further improvements is given.

High-Speed Silicon-Organic Hybrid Modulator: A modulator based on a silicon slot-waveguide and a second-order nonlinear cladding has been fabricated and tested. The 3 dB bandwidth surpasses – the first time for an integrated device – the psychological threshold of 100 GHz, opening the way to novel applications like radio-over-fiber. The modulation efficiency of $V_{\pi}L = 11.2$ Vmm is similar or in most cases smaller than for other high-speed modulators demonstrated in silicon so far.

The high device bandwidth has been obtained thanks to a gate voltage which diminished the RC time-constant well below 3 ps, without introducing significant optical loss. The use of such a gate voltage in an integrated optical device has been proposed here for the first time.

In the prototypes tested in this thesis, the gate voltage is applied across the 2 µm substrate; therefore the required electric-field strengths require voltages up to 400 V. The gate voltage can be diminished in future devices, as explained in Chapter 2, by using thin oxide, and doped polysilicon gates. Although the deposition of a gate-insulator and a gate is a standard process for certain types of CMOS processes, this has never been made so far in combination with optical waveguides, requiring to reconsider the process-flows.

Additionally, the driving voltage strongly depends from the nonlinearity achieved in the nanoslot. This depends mainly from the nonlinear material used, and from the poling process. It was evident from several experiments that the DC-nonlinearity appears higher than the nonlinearity measured at frequencies of 1 GHz or higher. Unfortunately, most –if not all- the values reported in the literature describe the nonlinearity in the low-frequency limit. The nonlinearity reached in this thesis for modulating frequencies larger than 1 GHz is the highest value ever reported in silicon-organic hybrid slot waveguides, but is still smaller than the value specified by the producer. It remains therefore to clarify if the poling procedure can be further optimized, or if this discrepancy is simply due to the frequency-dependent characteristic of the nonlinear material.

Second-order nonlinear silicon-organic hybrid waveguide: Based on numerical simulations and dispersion-engineering, a novel waveguide structure for phase-matched three-wave mixing has been analyzed. This is the first time that such a device has been proposed within the silicon-organic hybrid (SOH) platform. The nonlinear conversion efficiency predicted by using conventional nonlinear organic materials is one of the highest values that can be found in the literature. Such a device still remains to be experimentally tested, and might lead to the cheapest and most efficient method for generating and amplifying mid-IR wavelengths.

Liquid Crystal Phase Shifter on the SOH Platform with Ultra-Low Power Consumption: A silicon phase shifter with a record-low $V_{\pi}L = 0.085$ Vmm and power consumption in the nW range has been demonstrated. The bandwidth of the device is limited to few ms; which is enough for a number of low-speed applications such as optical matrix switches, polarization controllers, tunable filters and delay-interferometers. The device has been operated under normal atmosphere for time of the characterization. For obtaining a useful product, suited packages for containing the liquid-crystal remain to be built.

Appendix

Recipe for in-House Fabrication of Silicon Rib Waveguides

The rib waveguides used in this thesis for determining the optical absorption of inversion/accumulation layers have been fabricated in-house. Here we report on the process flow. The substrate used is a 10 × 14 mm SOI chip having a device layer 220 nm thick and a BOX thickness of 2 μm.

The chip was processed as follows:

Sample preparation:

- 5-30 s acetone (ACE) rinse for removing organic residues (e.g. resist).
- 5 s isopropanol (ISO) rinse for removing acetone residues.
- 5 s deionized (DI) rinse for removing isopropanol and possible particles. Higher water resistivity indicates lower ion concentration, and a stronger repulsive force between particles of the same material due to a stronger zeta potential.
- Nitrogen blow for removing remaining amounts of water and particles floating in it.
- 30 s oxygen plasma.
- 5 min hotplate at 150 °C for water desorption.

Spin coating

- Spin coating of the adhesion promoter Ti-prime (available from MicroChemicals GmbH), 4000 rpm, acceleration 1000 rpm/s, 30 s.
- Hotplate 120 °C for 2 min (1 mm thick, initially cold, aluminum plate under the chips) for activation of the Ti-prime. This transforms the organic molecules containing titanium in a non-closed atomic layer of TiO_2.
- Spin coating of the negative-tone resist ma-N 2401 (available from micro resist technology GmbH), 3000 rpm, 3000 rpm/s, 30 s. Nominal thickness: 100 nm.

Lithography

- e-beam lithography on a JEOL system, with 70 μC/cm^2 dose and 4 nm step size.

Development

- 12s in developer Dev-525 (available at micro chem GmbH). This value differs from what is specified in the datasheet; we attribute the discrepancy to a smaller focus size of our e-beam lithography system.
- 30 s immersion in a first beaker DI water.

  30 s immersion in a second beaker DI water (stopper).

Removal of the TiO_2 layer

- 2s in 25 % HF. This will remove the titanium oxide formed after activation of the adhesion promoter Ti-prime. This layer is highly resistant against SF_6 etching. According to the producer (private communication), even 1 s in 1 % HF would suffice.
- Rinse in DI water.
- The following reactive-ion etching step must be performed immediately afterwards for avoiding the formation of SiO_2.

Reactive-ion etching (RIE)

- 7 s in a Sentec Etchlab 200 RIE, SF_6, pressure 0.1 Pa, gas flow 5 sccm, RF power 500 W. This resulted in 70 nm etch depth. The low pressure is chosen for achieving a long gas mean free path length for higher kinetic energy of the ions (mean free path larger than the electrode separation). The RF power was chosen high for the same reason. This way a significant physical etching, as opposed to chemical etching, can rapidly remove any remaining contamination on the silicon surface without delaying the beginning of the silicon etch. The etch rate was determined immediately before by etching the back-side of an identical SOI chip for 2 h, and measuring the thickness difference with a micrometer. An etch rate of 10.0 nm/s was determined.

Fabrication of Electrodes

The silver electrodes of the 100 GHz modulator described in Section 2.3 have been deposited in house by means of a lift-off process on SOI chips structured by LETI. In this section we summarize the processing steps. After having determined the desired geometry by numerical simulations [28], the design was written in gds format by means of the Python-based package IPKISS [124]. The structure was written on a soda lime antireflection-coated 3×3×0.06 inch chromium mask by Compugraphics Jena GmbH (former ML&C) with 200 nm CD-tolerance and 100 nm write grid. The samples used had a size of 12×12 mm and contained four identical reticles. The large sample size was chosen in order to minimize the effects of a thick resist layer forming at the chip's edges by spin coating (edge bead).

The samples were processed as follows:

Sample preparation:

- 5-30 s acetone (ACE) rinse for removing organic residues (e.g. resist).
- 5 s isopropanol (ISO) rinse for removing acetone residues.
- 5 s deionized (DI) rinse for removing isopropanol and eventual particles.
- Nitrogen blow for removing remaining amounts of water and particles floating in it.
- 30 s oxygen plasma.
- 5 min hotplate at 150 °C for water desorption.

Spin coating

- Spin coating of the adhesion promoter Ti-prime (available from MicroChemicals GmbH), 4000 rpm, acceleration 1000 rpm/s, 30 s.
- Hotplate 120 °C for 2 min (1 mm thick, initially cold, aluminum plate under the chips) for activation of the Ti-prime. This transforms the organic molecules containing titanium in a non-closed atomic layer of TiO_2.
- Spin coating of the negative-tone resist ma-N 1410 (available from micro resist technology GmbH), 3000 rpm, 3000 rpm/s, 30 s followed by 6000 rpm, 3000 rpm/s, 2 s for reduction of the edge bead. Nominal thickness: 1 µm. The ma-N 1410 resist strongly absorbs UV light, resulting in a less intense illumination at the bottom of the resist layer, and consequently to non-vertical sidewalls which are ideal for lift-off.

Lithography

- Lithography for 47 s at 7.5 mW/cm^2 (dose of 350 mJ/cm^2). Wavelength 365 nm.

Development

- 15 s immersion in buffered hydrofluoric acid (BHF) for removing the estimated 10 nm thick thermal oxide layer produced during fabrication (SiO2 etch rate: 2 nm/s). This step is necessary for guaranteeing an ohmic contact between metal electrodes and silicon.
- 30 s immersion in a first beaker DI water.
- 30 s immersion in a second beaker DI water (stopper).

Metallization

- e-beam evaporation of 5 nm chromium. Sample must lie perfectly perpendicular to the source. Metal-source to sample distance used: 30 cm, vacuum level: 10^{-6} mBar.
- e-beam evaporation of 400 nm of silver (purity 99.99 %, silver slugs from Sigma-Aldrich).

Lift-off

- 24 h bath in acetone. Chips must be placed face down, e.g. on concave Petri plates (Uhrglassschalen).
- Removal of remaining metal using a gentle jet of acetone (a Pasteur pipette can be used for this). Use of ultrasound should be absolutely avoided for preventing scratches on the final structure.

Vias for Contacting Accumulation/Inversion Layers at Temperatures close to 0 °K

Since silicon is a semiconductor, the intrinsic carrier concentration diminishes dramatically with decreasing temperature. Even if silicon would become a perfect insulator only at 0 °K, when the temperature approaches the absolute zero, 1 mm^2 films as thin as 220 nm acquires a resistance so high that a four-point resistance measurement becomes impossible. To circumvent this problem, we fabricated nickel silicide (NiSi) contacts in the 220 nm thick silicon layer of our samples. Nickel silicide indeed is metallic [125], and therefore remains conductive even at 0 °K.

The recipe used is derived from [126] and is given below:

- A stainless steel mask having 1.5 mm holes at the desired position is fixed 0.1 mm above the silicon surface.
- The sample is placed in a vacuum chamber with argon pressure of approximately 10^{-2} mbar.
- The sample is heated to 600 °C.
- 1 nm of Ni is sputtered on the sample. This should crack the native oxide on the silicon surface.
- The sample is cooled to 400 °C.
- 400 nm of Ni are sputtered in 30 min. Mono nickel silicide is formed.
- The sample is cooled to room temperature.

Samples created in such a way maintained electrical contact with the inversion/accumulation layers formed at the Si/SiO_2 interface even at temperatures as low as 40 mK, Fig. 28.

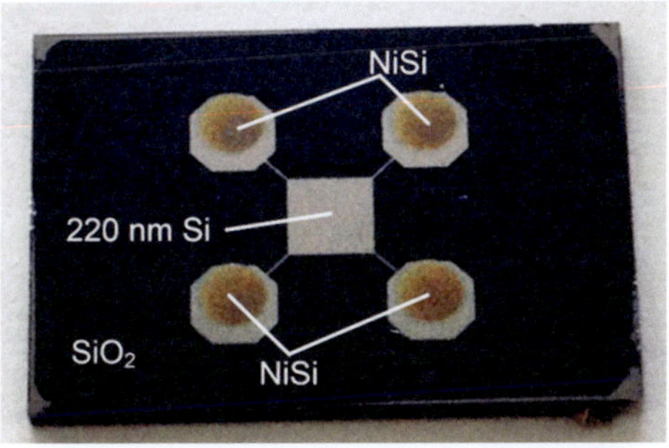

Fig. 28. Photograph of an SOI chip with four NiSi contacts in the 220 nm thick device silicon layer. The silicon dioxide is visible where the silicon has been etched (darker region).

Impedance and Propagation Loss from a Vector Network Analyzer (VNA) Measurement

In this section we present the algorithm used for extracting the impedance and the propagation loss of an electrical waveguide, such as the coplanar line of an electro-optic modulator. The model used is based on transmission line model, where currents and electrical potentials at every position of the line are well defined. This is however an approximation, which is valid only in the low-frequency limit. Indeed, when a fast varying magnetic field $\boldsymbol{B}$ is present, the third Maxwell equation rot$\boldsymbol{E}$ = -d$\boldsymbol{B}$/dt, shows that the electrical integral along a closed line is generally different from zero. At the frequencies and the device-sizes involved, only an integration of Maxwell equations, by means of a numerical solver for example, can provide exact solutions. The following treatment however still provides a good insight to the problem.

The model considered here is shown in Fig. 29. The device under test (DUT) have unknown impedance Z_{DUT} and $\gamma = -i\beta+\alpha/2$ where β is the propagation constant and α is the propagation loss (both real). The device length L is assumed to be known.

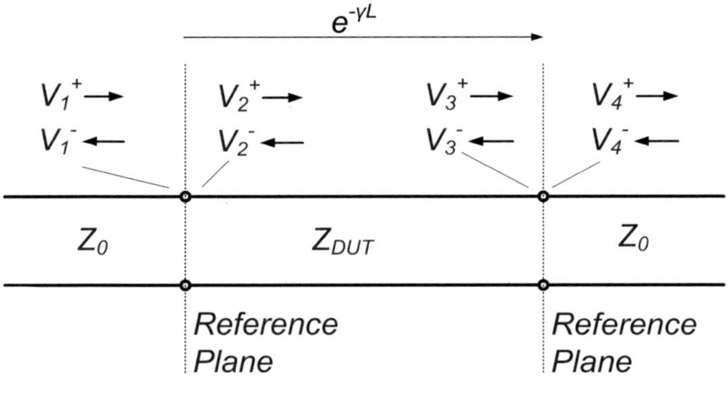

Fig. 29. Transmission line model. A device having impedance Z_{DUT}, and length L is connected on two sides with lines having impedance Z_0 (for example the standard 50 Ω). The Z_0 lines represent the cable of the VNA, which is calibrated on the reference planes. Sinusoidal waves propagating from left to right have a voltage amplitude V_i^+ at the positions i = 1, 2, 3, 4. Waves propagating in the opposite direction are indicated with V_i^-. When the waves in the DUT propagate from left to right they acquire an amplitude $e^{-\gamma}$, where γ is a complex number.

On the left reference plane in Fig. 29, equating voltage and current on the left side to the values on the right side gives

$$\begin{cases} V_1^+ + V_1^- = V_2^+ + V_2^- \\ \dfrac{1}{Z_0}\left(V_1^+ - V_1^-\right) = \dfrac{1}{Z_{DUT}}\left(V_2^+ - V_2^-\right). \end{cases} \tag{A.1}$$

The latter equations can be rewritten in matrix form:

$$\begin{pmatrix} V_2^+ \\ V_2^- \end{pmatrix} = \frac{1}{2Z_0} \begin{pmatrix} Z_0 + Z_{DUT} & Z_0 - Z_{DUT} \\ Z_0 - Z_{DUT} & Z_0 + Z_{DUT} \end{pmatrix} \begin{pmatrix} V_1^+ \\ V_1^- \end{pmatrix} \equiv M_{21} \begin{pmatrix} V_1^+ \\ V_1^- \end{pmatrix} \tag{A.2}$$

The propagation across the DUT can be described by:

$$\begin{pmatrix} V_3^+ \\ V_3^- \end{pmatrix} = \begin{pmatrix} e^{-\gamma L} & 0 \\ 0 & e^{\gamma L} \end{pmatrix} \begin{pmatrix} V_2^+ \\ V_2^- \end{pmatrix} \equiv M_{32} \begin{pmatrix} V_2^+ \\ V_2^- \end{pmatrix} \tag{A.3}$$

Finally the relationship between the amplitudes on the second reference plane can be written analogously to Eq. (A.2):

$$\begin{pmatrix} V_4^+ \\ V_4^- \end{pmatrix} = \frac{1}{2Z_{DUT}} \begin{pmatrix} Z_{DUT} + Z_0 & Z_{DUT} - Z_0 \\ Z_{DUT} - Z_0 & Z_{DUT} + Z_0 \end{pmatrix} \begin{pmatrix} V_3^+ \\ V_3^- \end{pmatrix} \equiv M_{43} \begin{pmatrix} V_3^+ \\ V_3^- \end{pmatrix}. \tag{A.4}$$

It follows immediately:

$$\begin{pmatrix} V_4^+ \\ V_4^- \end{pmatrix} = M_{43} M_{32} M_{21} \begin{pmatrix} V_1^+ \\ V_1^- \end{pmatrix}. \tag{A.5}$$

The transmission coefficient T (from right to left) and the reflection coefficient R can be obtained by setting $V_1^+ = 0$ in the latter equation:

$$T \equiv \frac{V_1^-}{V_4^-} = \frac{4Z_{DUT} Z_0}{e^{\gamma L}(Z_0 + Z_{DUT})^2 - e^{-\gamma L}(Z_0 - Z_{DUT})^2} = \frac{(1-\Gamma^2)e^{-\gamma L}}{1-\Gamma^2 e^{-2\gamma L}} \tag{A.6}$$

$$R \equiv \frac{V_4^+}{V_4^-} = \frac{(Z_0^2 - Z_{DUT}^2)(e^{-\gamma L} - e^{+\gamma L})}{e^{\gamma L}(Z_0 + Z_{DUT})^2 - e^{-\gamma L}(Z_0 - Z_{DUT})^2} = \frac{\Gamma(1-e^{-2\gamma L})}{1-\Gamma^2 e^{-2\gamma L}} \tag{A.7}$$

where:

$$\Gamma \equiv \frac{Z_{DUT} - Z_0}{Z_{DUT} + Z_0} \tag{A.8}$$

$$\gamma \equiv -i\beta + \frac{\alpha}{2}. \tag{A.9}$$

Equations (A.6) and (A.9) can be inverted, allowing to calculate Γ and γ in terms of T and R:

$$\Gamma = \frac{1 + R^2 - T^2 - \sqrt{-4T^2 + (R^2 - T^2 - 1)^2}}{2R} \tag{A.10}$$

$$e^{-\gamma L} = \frac{1 - R^2 + T^2 - \sqrt{-4T^2 + (R^2 - T^2 - 1)^2}}{2T} \tag{A.11}$$

The device impedance, propagation constant and propagation loss can be extracted using Eqs. (A.10) and (A.11) from the S-parameters which can be measured with a Vector Network Analyzer (VNA).

Matlab Code for Numerical Integration of MIS Equations

The code below was used for integrating Eq. (1.53) when a gate voltage is applied on the substrate of an SOI wafer. The code calculates for a given voltage the electric field at the Si/SiO_2 interface (silicon side), Eq. is solved numerically Eq. (1.53) providing the surface potential. This value defines the boundary conditions for the differential equation (1.53), which is integrated numerically.

```
clear all;
close all;

% Constants for the MIS model:
ni = 1.02e10; % Intrinsic carriers concentration of Si. 1/cm3
ni = ni * 1e6; % Intrinsic carriers concentration of Si. 1/m3
q = 1.6e-19; % elementary charge. Coulomb
kB = 1.38e-23; % m2 kg/s2/k
T = 300; % temperature in K.
eps0 = 8.85e-12; %Free space permittivity, F/m.
eps_Si = 11.7; % Relative permittivity, Silicon.
eps_BOX = 3.8; % Relative permittivity, BOX.
d_BOX = 2e-6; % thickness BOX in m.
x_max_integration = 200e-9; %m.

% Wafer type, bulk ion concentration:      ****** PARAMETERS *******
n0 = 1e15; % Equilibrium electron density. 1/cm3
n0 = n0 * 1e6;  % Equilibrium electron density. 1/m3
p0 = ni^2/n0; % Equilibrium hole density. 1/m3
%
beta = q/kB/T;
% Defining the function relating electric field and potential psi:
E = @(psi, n0) sign(psi).* (2.*kB.*T.*n0./eps0./eps_Si).^0.5 ...
    .* ( (exp(beta.*psi) - beta.*psi - 1) ...
    + (ni./n0)^2 .* (exp(-beta.*psi) + beta.*psi - 1) ).^0.5;
%
% Given and E_desired, finding psi which creates it.
i_gate = 1; % index of gate voltage.
for V_gate = cat(2,-500:10:-10 , 10:10:500) % list of gate voltages.
E_BOX = V_gate/d_BOX;
E_Si = E_BOX*eps_BOX/eps_Si;
E_desired =E_Si;
to_be_zero = @(psi, n0) E(psi, n0) - E_desired;

psi_old = 0.1*sign(E_desired); % initial value.
d_psi = 0.01; % initial step.
while ( d_psi > 1e-15 ) % Using Newton's method to find the root.
D_to_be_zero = ( to_be_zero(psi_old + d_psi, n0) ...
               - to_be_zero(psi_old - d_psi, n0) )/(2*d_psi); % Derivative
of the function "to_be_zero".
psi_new = psi_old - 0.01*to_be_zero(psi_old, n0)/D_to_be_zero;
d_psi = abs(psi_new - psi_old)/100; % making sure the step is small enough.
psi_old = psi_new;
end

psi_s = psi_new; % Potential on the surface.
E(psi_s, n0)
```

```
if ( E(psi_s, n0) - E_desired )/E_desired < 0.01 % Check that the electric field obtained is within 1% accuracy.
    OK = 1
else
    ERROR =1
end
ns = n0 * exp(beta*psi_s); % surface density of electrons
ps = ni^2/n0 * exp(-beta*psi_s); % surface density of holes

% Integrating the potential psi from the surface to the bulk. i.e. from
% left to right, from x = 0 to x = infty.

x = 0; % in m.
psi_old = psi_s; % surface potential as initial value
i_x = 1; % index of position x.
while(x < x_max_integration)
    d_x = 1e-4* abs( psi_old / E(psi_old,n0) );
    psi_new = psi_old -E(psi_old, n0) * d_x;
    x = x + d_x;
    n = n0 * exp(beta*psi_new); % 1/m3
    p = ni^2/n0 * exp(-beta*psi_new); % 1/m3

    out_x(i_x, i_gate) = x;
    out_psi(i_x, i_gate) = psi_new;
    out_n(i_x, i_gate) = n;  % 1/m3
    out_p(i_x, i_gate) = p;  % 1/m3
    psi_old = psi_new;
    i_x = i_x + 1;
end
i_gate = i_gate +1;
end % end loop on i_gate.
```

Sellmeier Equations for Crystalline Silicon and Silicon Dioxide

In this appendix we report the dispersion equations used to model both the silicon and the silicon-dioxide of the device we fabricated. In all devices described in this thesis, silicon-on-insulator wafers produced by SOITEC have been used. The silicon constituting the optical waveguides (device layer) is crystalline, and the buried silicon-oxide (BOX) is thermally-grown oxide on a crystalline silicon wafer. For the device silicon layer, we assume that the dispersion of bulk silicon applies (neglecting for example possible mechanical tensions present in this layer). We use therefore the following equation which is valid up to 0.1 % in the wavelength range 1.12 - 588 µm [93]

$$n_{\mathrm{Si}}(\lambda)=3.41906+\frac{0.123172}{\lambda^2-0.028}+\frac{0.0265456}{\left(\lambda^2-0.028\right)^2}-2.66511\times10^{-8}\lambda^2+5.45852\times10^{-14}\lambda^4 \quad \text{(A.12)}$$

where λ is expressed in µm.

The dispersion of thermally-grown oxide on crystalline silicon has been studied by Herzinger et al. by ellipsometry [92] in the range 0.75 - 6.5 eV (1653 - 191 nm). In the measured range, it was found a variation of less than 1 % from the values calculated with the Sellmeier equations published by Palik for glaseous SiO_2 [93]. Additionally, a corrected Sellmeier equation was provided. However, since its validity range does not cover the mid-IR wavelengths, and because the correction introduced is very small, we use the equation of Malitson [127] which is valid in the range 0.21 - 3.71 µm:

$$n_{\mathrm{SiO_2}}(\lambda)=1+\frac{0.6961663\lambda^2}{\lambda^2-0.0684043^2}+\frac{0.4079426\lambda^2}{\lambda^2-0.1162414^2}+\frac{0.8974794\lambda^2}{\lambda^2-9.896161^2} \quad \text{(A.13)}$$

where λ is expressed in µm. A table of refractive index values is given in the following section.

Refractive Indices of Si and SiO_2

For the reader's convenience we report here a conversion table between different common units used to measure the photon's energy in the mid-IR and near-IR range, together with the refractive indices of the silicon and the silicon-dioxide described in the previous section.

Vacuum Wavelength [µm]	Vacuum Wavenumber [1/cm]	Frequency [THz]	Energy [eV]	Refractive Index Si	Refractive Index SiO_2
0.500	20000.00	600.00	2.479		1.4623
0.600	16666.67	500.00	2.066		1.4580
0.700	14285.71	428.57	1.771		1.4553
0.800	12500.00	375.00	1.550		1.4533
0.900	11111.11	333.33	1.377		1.4518
1.000	10000.00	300.00	1.240		1.4504
1.100	9090.91	272.73	1.127	3.5423	1.4492
1.200	8333.33	250.00	1.033	3.5196	1.4481
1.300	7692.31	230.77	0.954	3.5028	1.4469
1.400	7142.86	214.29	0.886	3.4899	1.4458
1.500	6666.67	200.00	0.826	3.4799	1.4446
1.550	6451.61	193.55	0.800	3.4756	1.4440
1.600	6250.00	187.50	0.775	3.4718	1.4434
1.700	5882.35	176.47	0.729	3.4653	1.4422
1.800	5555.56	166.67	0.689	3.4600	1.4409
1.900	5263.16	157.89	0.652	3.4555	1.4395
2.000	5000.00	150.00	0.620	3.4518	1.4381
2.100	4761.90	142.86	0.590	3.4486	1.4366
2.200	4545.45	136.36	0.564	3.4458	1.4350
2.300	4347.83	130.43	0.539	3.4434	1.4334
2.400	4166.67	125.00	0.517	3.4414	1.4316
2.500	4000.00	120.00	0.496	3.4395	1.4298
2.600	3846.15	115.38	0.477	3.4379	1.4279
2.700	3703.70	111.11	0.459	3.4365	1.4259
2.800	3571.43	107.14	0.443	3.4353	1.4238
2.900	3448.28	103.45	0.427	3.4341	1.4216
3.000	3333.33	100.00	0.413	3.4331	1.4192
3.100	3225.81	96.77	0.400	3.4322	1.4168
3.200	3125.00	93.75	0.387	3.4314	1.4143
3.300	3030.30	90.91	0.376	3.4306	1.4116
3.400	2941.18	88.24	0.365	3.4299	1.4088
3.500	2857.14	85.71	0.354	3.4293	1.4059
3.600	2777.78	83.33	0.344	3.4287	1.4028
3.700	2702.70	81.08	0.335	3.4282	1.3996
3.800	2631.58	78.95	0.326	3.4277	
3.900	2564.10	76.92	0.318	3.4273	
4.000	2500.00	75.00	0.310	3.4269	
4.200	2380.95	71.43	0.295	3.4261	
4.400	2272.73	68.18	0.282	3.4255	
4.600	2173.91	65.22	0.270	3.4249	
4.800	2083.33	62.50	0.258	3.4245	
5.000	2000.00	60.00	0.248	3.4240	
5.500	1818.18	54.55	0.225	3.4232	
6.000	1666.67	50.00	0.207	3.4225	
6.500	1538.46	46.15	0.191	3.4220	
7.000	1428.57	42.86	0.177	3.4216	
7.500	1333.33	40.00	0.165	3.4213	
8.000	1250.00	37.50	0.155	3.4210	
10.000	1000.00	30.00	0.124	3.4203	
12.000	833.33	25.00	0.103	3.4199	
14.000	714.29	21.43	0.089	3.4197	
16.000	625.00	18.75	0.077	3.4195	
18.000	555.56	16.67	0.069	3.4194	
20.000	500.00	15.00	0.062	3.4194	

Table of Symbols

Calligraphic Symbols

$\mathcal{E}_{gate}$ gate field, unit V/m, Section 1.3.2.

$\mathcal{E}_i^{+(-)}$ *i*-th component of the electric field of a mode propagating in the positive (negative) direction, unit V/m, Eq. (1.25).

$\mathcal{H}_i^{+(-)}$ *i*-th component of the magnetic field of a mode propagating in the positive (negative) direction, unit A/m, Eq. (1.26).

Latin Symbols

$A^{(v)}$ mode amplitude, dimensionless, Eq. (1.25).

cc. complex conjugate.

c speed of light, unit m/s, Eq. (1.43).

D_i *i*-th component of the displacement field, unit C/m^2, Eq. (1.1).

$\mathbf{E}$ electric field, unit V/m, Eq. (1.23).

E_F Fermi energy, unit J, Eq. (1.46).

$E_{c(v)}$ conduction (valence) band edge, unit J, Section 1.3.

$\mathbf{H}$ magnetic field, unit A/m, Eq. (1.23).

$\mathbf{J}$ electrical source, unit $C/(sm^2)$, Eq. (1.23).

k_B Boltzmann constant, unit J/K, Eq. (1.46).

$\mathbf{M}$ magnetic source, unit V/m^2, Eq. (1.23).

N_A acceptor density, unit $1/m^3$, Eq. (1.51).

N_D donor density, unit $1/m^3$, Eq. (1.51).

N_e electron density, unit $1/m^3$, Eq. (1.51).

N_h hole density, unit $1/m^3$, Eq. (1.51).

n_{eff} mode effective index, dimensionless, Eq. (3.1).

P_i *i*-th component of the electrical polarization density, units C/m^2, Eq. (1.30).

q elementary charge, unit C, Eq. (1.51).

r_{ijk} electro-optic tensor, unit m/V, Eq. (1.5).

r_{ij} electro-optic matrix, unit m/V, Eq. (1.15).

R_s sheet resistance, unit Ω/sq, Chapter 5.

R resistance, unit Ω, Chapter 5.

R reflection coefficient, dimensionless, Eq. (A.7).

T transmission coefficient, dimensionless, Eq. (A.6).

T temperature, unit K, Eq. (1.46).

V_π half-wavelength voltage, unit V, Eq. (1.21).

V_{gate} gate voltage, unit V, Eq. (1.54).

Z_0 reference line impedance, unit Ω, Eq. (A.1).

Z_{DUT} reference of the device-under-test, unit Ω, Eq. (A.1).

Greek Symbols

α_{wg} waveguide attenuation coefficient, unit 1/m, Eq. (1.44).

α local attenuation coefficient, unit 1/m, Eq. (5.1).

β mode propagation constant, unit 1/m, Eq. (1.39).

β_{TPA} two-photon absorption coefficient, unit m/W, Section 3.5.

Γ electro-optic interaction factor, dimensionless, Eq. (1.20).

ε_0 free-space permittivity, unit C/(Vm), Eq. (1.1).

ε_{ij} permittivity tensor, dimensionless, Eq. (1.1).

ε_s full scalar permittivity, unit C/(Vm), Eq. (1.40).

η_{ij} impermeability tensor, dimensionless, Eq. (1.2).

η normalized conversion efficiency, unit $1/(\text{Wm}^2)$, Eq. (3.12).

λ_0 free-space wavelength, unit m, Eq. (1.21).

ρ electrical resistivity, unit Ωm, Chapter 5.

$\chi^{(n)}$ n-th order susceptibility tensor, unit $(\text{m/V})^{(n-1)}$, Eq. (1.30).

ψ electric potential, unit V, Eq. (1.48).

ω angular frequency, unit 1/s, Eq. (1.22).

Bibliography

[1] C. Gunn, "Fully Integrated VLSI CMOS and Photonics "CMOS Photonics"," VLSI Technology, 2007 IEEE Symposium on, 6-9 (2007).

[2] L. Kimerling, "Scaling Information Technology on the Silicon Platform," European Conference on Optical Communication (ECOC 2010), paper Tu.4.C.1 (2010).

[3] L. Alloatti, D. Korn, R. Palmer, D. Hillerkuss, J. Li, A. Barklund, R. Dinu, J. Wieland, M. Fournier, J. Fedeli, H. Yu, W. Bogaerts, P. Dumon, R. Baets, C. Koos, W. Freude, and J. Leuthold, "42.7 Gbit/s electro-optic modulator in silicon technology," Optics Express **19**, 11841-11851 (2011).

[4] R. Ding, T. Baehr-Jones, Y. Liu, R. Bojko, J. Witzens, S. Huang, J. Luo, S. Benight, P. Sullivan, J. M. Fedeli, M. Fournier, L. Dalton, A. Jen, and M. Hochberg, "Demonstration of a low V pi L modulator with GHz bandwidth based on electro-optic polymer-clad silicon slot waveguides," Optics Express **18**, 15618-15623 (2010).

[5] L. Alloatti, R. Palmer, S. Diebold, P. Pahl, M. Pauli, J. Li, B. Chen, R. Dinu, M. Fournier, J. Fedeli, T. Zwick, C. Koos, W. Freude, and J. Leuthold, "100 GHz silicon-organic hybrid modulator," Submitted for publication.

[6] L. Alloatti, D. Korn, C. Weimann, C. Koos, W. Freude, and J. Leuthold, "Second-order nonlinear silicon-organic hybrid waveguides," Optics Express **20**, 20506-20515 (2012).

[7] L. Alloatti, J. Pfeifle, J. Mendez, W. Freude, L. J, and C. Koos, "Liquid crystal phase shifter on the SOH platform with ultra-low power consumption," Optical Fiber Communication Conference (OFC'12), paper OTu1I.5 (2012).

[8] J. Pfeifle, L. Alloatti, W. Freude, J. Leuthold, and C. Koos, "Silicon-organic hybrid phase shifter based on a slot waveguide with a liquid-crystal cladding," Optics Express **20**, 15359-15376 (2012).

[9] L. Alloatti, M. Lauermann, C. Suergers, C. Koos, W. Freude, and J. Leuthold, "Optical absorption in silicon layers in the presence of charge inversion/accumulation or ion implantation," Submitted for publication.

[10] L. Alloatti and et al., "Low-temperature sheet resistance of accumulation/inversion layers in silicon," In preparation.

[11] R. Palmer, L. Alloatti, D. Korn, M. Moosmann, K. Huska, U. Lemmer, D. Gerthsen, T. Schimmel, W. Freude, K. C, and L. J, "Smooth and ultra-precise silicon nanowires fabricated by conventional optical lithography," CLEO, paper CThZ1 (2011).

[12] R. Palmer, "A new method to fabricate nano-wire waveguides," Diploma thesis, IPQ, Karlsruhe Institute of Technology (KIT) (2010).

[13] C. Weimann, "New passive silicon on insulator devices," Diploma thesis, IPQ, Karlsruhe Institute of Technology (KIT) (2011).

[14] L. Alloatti, D. Korn, D. Hillerkuss, T. Vallaitis, J. Li, R. Bonk, R. Palmer, T. Schellinger, A. Barklund, R. Dinu, J. Wieland, M. Fournier, J. Fedeli, W. Bogaerts, P. Dumon, R. Baets, C. Koos, W. Freude, and J. Leuthold, "Silicon High-Speed Electro-Optic Modulator," Proceedings of IEEE Conference on Group IV Photonics, paper ThC2 (2010).

[15] R. Palmer, L. Alloatti, D. Korn, W. Heni, P. Schindler, J. Bolten, M. Karl, M. Waldow, T. Wahlbrink, W. Freude, C. Koos, and J. Leuthold, "Highly Efficient Strip-to-Slot Mode Converters," CLEO 2012, paper CM4M1 (2012).

[16] M. Boerner, "Mehrstufiges Uebertragungssytem fuer in Pulscodemodulation dargestellte Nachrichten, DE," DE 000 001 254 513 B (1966).

[17] Cisco, "http://www.cisco.com/en/US/solutions/collateral/ns341/ns525/ns537/ns705/ns827/white_paper_c11-481360_ns827_Networking_Solutions_White_Paper.html."

[18] L. Liao, A. Liu, D. Rubin, J. Basak, Y. Chetrit, H. Nguyen, R. Cohen, N. Izhaky, and M. Paniccia, "40 Gbit/s silicon optical modulator for high-speed applications," Electronics Letters **43**, 1196-1197 (2007).

[19] R. W. Boyd, *Nonlinear Optics* (Academic press, 2008).

[20] M. Born and E. Wolf, *Principles of optics* (Cambridge university press, 1983).

[21] L. R. Dalton, S. J. Benight, L. E. Johnson, D. B. Knorr, I. Kosilkin, B. E. Eichinger, B. H. Robinson, A. K. Y. Jen, and R. M. Overney, "Systematic Nanoengineering of Soft Matter Organic Electro-optic Materials," Chemistry of Materials **23**, 430-445 (2011).

[22] Y. Enami, C. T. Derose, D. Mathine, C. Loychik, C. Greenlee, R. A. Norwood, T. D. Kim, J. Luo, Y. Tian, A. K. Y. Jen, and N. Peyghambarian, "Hybrid polymer/sol-gel waveguide modulators with exceptionally large electro-optic coefficients," Nature Photonics **1**, 180-185 (2007).

[23] Agilent, "http://cp.literature.agilent.com/litweb/pdf/5989-7808EN.pdf."

[24] A. Liu, L. Liao, D. Rubin, J. Basak, Y. Chetrit, H. Nguyen, R. Cohen, N. Izhaky, and M. Paniccia, "Recent development in a high-speed silicon optical modulator based on reverse-biased pn diode in a silicon waveguide," Semiconductor Science and Technology **23**, 064001-064001 (2008).

[25] Fujitsu, "http://jp.fujitsu.com/group/foc/en/services/optical-devices/40gln."

[26] J. Witzens, T. Baehr-Jones, and M. Hochberg, "Design of transmission line driven slot waveguide Mach-Zehnder interferometers and application to analog optical links," Optics Express **18**, 16902-16027 (2010).

[27] J. M. Brosi, C. Koos, L. C. Andreani, M. Waldow, J. Leuthold, and W. Freude, "High-speed low-voltage electro-optic modulator with a polymer-infiltrated silicon photonic crystal waveguide," Optics Express **16**, 4177-4191 (2008).

[28] Cst, "http://www.cst.com/Content/Products/MWS/Overview.aspx."

[29] A. Melikyan, "RSoft simulations for second order nonlinear electro-optical modulators with different geometrical properties," Bachelor thesis, IPQ, Karlsruhe Institute of Technology (KIT) (2010).

[30] C. Vassallo, *Optical Waveguide Concepts* (Elsevier, 1991).

[31] J. D. Jackson, *Classical Electrodynamics* (John Wiley & Sons Ltd., 1962).

[32] J. C. Stang, "100 GHz silicon transistors," Diploma thesis, IPQ, Karlsruhe Institute of Technology (KIT) (2012).

[33] S. M. Sze and K. Ng Kwok, *Physics of Semiconductor Devices, Third Edition* (Wiley, 2006).

[34] S. P. Voinigescu, T. O. Dickson, R. Beerkens, I. Khalid, and P. Westergaard, "A comparison of SiCMOS, SiGeBiCMOS, and InPHBT technologies for high-speed and millimeter-wave ICs," 2004 Topical Meeting on Silicon Monolithic Integrated Circuits inRF Systems, Digest of Papers, 111-114 (2004).

[35] R. S. Jacobsen, K. N. Andersen, P. I. Borel, J. Fage-Pedersen, L. H. Frandsen, O. Hansen, M. Kristensen, A. V. Lavrinenko, G. Moulin, H. Ou, C. Peucheret, B. Zsigri, and A. Bjarklev, "Strained silicon as a new electro-optic material," Nature **441**, 199-202 (2006).

[36] P. Dong, S. R. Liao, D. Z. Feng, H. Liang, D. W. Zheng, R. Shafiiha, C. C. Kung, W. Qian, G. L. Li, X. Z. Zheng, A. V. Krishnamoorthy, and M. Asghari, "Low V-pp, ultralow-energy, compact, high-speed silicon electro-optic modulator," Optics Express **17**, 22484-22490 (2009).

[37] W. M. J. Green, M. J. Rooks, L. Sekaric, and Y. A. Vlasov, "Ultra-compact, low RF power, 10 gb/s silicon Mach-Zehnder modulator," Optics Express **15**, 17106-17113 (2007).

[38] G. T. Reed, G. Mashanovich, F. Y. Gardes, and D. J. Thomson, "Silicon optical modulators," Nature Photonics **4**, 518-526 (2010).

[39] T. W. Baehr-Jones and M. J. Hochberg, "Polymer silicon hybrid systems: A platform for practical nonlinear optics," Journal of Physical Chemistry C **112**, 8085-8090 (2008).

[40] M. Hochberg, T. Baehr-Jones, G. Wang, J. Huang, P. Sullivan, L. Dalton, and A. Scherer, "Towards a millivolt optical modulator with nano-slot waveguides," Optics Express **15**, 8401-8410 (2007).

[41] J. H. Wülbern, A. Petrov, and M. Eich, "Electro-optical modulator in a polymer-infiltrated silicon slotted photonic crystal waveguide heterostructure resonator," Optics Express **17**, 304-313 (2009).

[42] C. Koos, J. Brosi, M. Waldow, W. Freude, and J. Leuthold, "Silicon-on-insulator modulators for next-generation 100 Gbit/s-Ethernet," Proceedings of European Conference on Optical Communication (ECOC), P056-P056 (2007).

[43] W. Freude, J. Leuthold, L. Alloatti, T. Vallaitis, D. Korn, R. Palmer, C. Koos, J. M. Brosi, P. Dumon, R. Baets, M. L. Scimeca, I. Biaggio, B. Breiten, F. Diederich, A. Barklund, R. Dinu, and J. Wieland, "100 Gbit/s electro-optic modulator and 56 Gbit/s wavelength converter for DQPSK data in silicon-organic hybrid (SOH) technology," Proceedings of IEEE Conference on Novel Waveguiding, Structures and Phenomena, Summer Topicals, paper WB2.1 (2010).

[44] J. H. Wülbern, S. Prorok, J. Hampe, A. Petrov, M. Eich, J. D. Luo, A. K. Y. Jen, M. Jenett, and A. Jacob, "40 GHz electro-optic modulation in hybrid silicon-organic slotted photonic crystal waveguides," Optics Letters **35**, 2753-2755 (2010).

[45] T. Baehr-Jones, B. Penkov, J. Q. Huang, P. Sullivan, J. Davies, J. Takayesu, J. D. Luo, T. D. Kim, L. Dalton, A. Jen, M. Hochberg, and A. Scherer, "Nonlinear polymer-clad silicon slot waveguide modulator with a half wave voltage of 0.25 V," Applied Physics Letters **92**, 163303-163303 (2008).

[46] J. Leuthold, W. Freude, J. M. Brosi, R. Baets, P. Dumon, I. Biaggio, M. L. Scimeca, F. Diederich, B. Frank, and C. Koos, "Silicon Organic Hybrid Technology-a Platform for Practical Nonlinear Optics," Proceedings of the IEEE **97**, 1304-1316 (2009).

[47] Q. Lin, O. J. Painter, and G. P. Agrawal, "Nonlinear optical phenomena in silicon waveguides: Modeling and applications," Optics Express **15**, 16604-16644 (2007).

[48] N. N. Feng, S. R. Liao, D. Z. Feng, P. Dong, D. W. Zheng, H. Liang, R. Shafiiha, G. L. Li, J. E. Cunningham, A. V. Krishnamoorthy, and M. Asghari, "High speed carrier-depletion modulators with 1.4V-cm V pi L integrated on 0.25 mu m silicon-on-insulator waveguides," Optics Express **18**, 7994-7999 (2010).

[49] J. S. Suehle and P. Chaparala, "Low electric field breakdown of thin SiO2 films under static and dynamic stress," IEEE Transactions on Electron Devices **44**, 801-808 (1997).

[50] GigOptix, "www.gigoptix.com."

[51] D. Jin, H. Chen, A. Barklund, J. Mallari, G. M. Yu, E. Miller, and R. Dinu, "EO Polymer Modulators Reliability Study," Organic Photonic Materials and Devices XII **7599**, 75990H (2010).

[52] Epixfab, "http://www.epixfab.eu/design/predefined_masks/imec/fxt_vertical_fiber_couplers."

[53] J. Blasco and C. A. Barrios, "Compact slot-waveguide/channel-waveguide mode-converter," Lasers and Electro-Optics Europe, 2005. CLEO/Europe. 2005 Conference on, 607 (2005).

[54] C. Gunn, "CMOS photonics for high-speed interconnects," IEEE Micro **26**, 58-66 (2006).

[55] C. Koos, P. Vorreau, T. Vallaitis, P. Dumon, W. Bogaerts, R. Baets, B. Esembeson, I. Biaggio, T. Michinobu, F. Diederich, W. Freude, and J. Leuthold, "All-optical high-speed signal processing with silicon-organic hybrid slot waveguides," Nature Photonics **3**, 216-219 (2009).

[56] Synopsys, "http://www.synopsys.com/Tools/TCAD/DeviceSimulation/Pages/default.aspx."

[57] X. Z. Zheng, J. Lexau, Y. Luo, H. Thacker, T. Pinguet, A. Mekis, G. L. Li, J. Shi, P. Amberg, N. Pinckney, K. Raj, R. Ho, J. E. Cunningham, and A. V. Krishnamoorthy, "Ultra-low-energy all-CMOS modulator integrated with driver," Optics Express **18**, 3059-3070 (2010).

[58] S. S. Li and W. R. Thurber, "Dopant Density and Temperature-dependence of Electron-mobility and Resistivity in *n*-type Silicon," Solid-State Electronics **20**, 609-616 (1977).

[59] R. A. Soref and B. R. Bennett, "Electrooptical Effects in Silicon," IEEE Journal of Quantum Electronics **23**, 123-129 (1987).

[60] D. M. Caughey and R. E. Thomas, "Carrier Mobilities in Silicon Empirically Related to Doping and Field," Proceedings of the Institute of Electrical and Electronics Engineers **55**, 2192-2193 (1967).

[61] W. Spitzer and H. Y. Fan, "Infrared Absorption in *n*-Type Silicon," Physical Review **108**, 268-271 (1957).

[62] Leti, "http://www.epixfab.eu/mpw_shuttles/shuttles/leti03."

[63] Z. W. Shi, J. D. Luo, S. Huang, B. M. Polishak, X. H. Zhou, S. Liff, T. R. Younkin, B. A. Block, and A. K. Y. Jen, "Achieving excellent electro-optic activity and thermal stability in poled polymers through an expeditious crosslinking process," Journal of Materials Chemistry **22**, 951-959 (2012).

[64] X. S. Yao and L. Maleki, "Optoelectronic microwave oscillator," Journal of the Optical Society of America B-optical Physics **13**, 1725-1735 (1996).

[65] I. L. Gheorma and G. K. Gopalakrishnan, "Flat frequency comb generation with an integrated dual-parallel modulator," IEEE Photonics Technology Letters **19**, 1011-1013 (2007).

[66] T. Baehr-Jones, R. Ding, Y. Liu, A. Ayazi, T. Pinguet, N. C. Harris, M. Streshinsky, P. Lee, Y. Zhang, A. E. J. Lim, T. Y. Liow, S. H. G. Teo, G. Q. Lo, and M. Hochberg, "Ultralow drive voltage silicon traveling-wave modulator," Optics Express **20**, 12014-12020 (2012).

[67] D. J. Thomson, F. Y. Gardes, J. M. Fedeli, S. Zlatanovic, Y. F. Hu, B. P. P. Kuo, E. Myslivets, N. Alic, S. Radic, G. Z. Mashanovich, and G. T. Reed, "50-Gb/s Silicon Optical Modulator," IEEE Photonics Technology Letters **24**(2012).

[68] Y. B. Tang, J. D. Peters, and J. E. Bowers, "Over 67 GHz bandwidth hybrid silicon electroabsorption modulator with asymmetric segmented electrode for 1.3 mu m transmission," Optics Express **20**, 11529-11535 (2012).

[69] R. Dinu, D. Jin, G. M. Yu, B. Q. Chen, D. Y. Huang, H. Chen, A. Barklund, E. Miller, C. L. Wei, and J. Vemagiri, "Environmental Stress Testing of Electro-Optic Polymer Modulators," Journal of Lightwave Technology **27**, 1527-1532 (2009).

[70] M. Ebrahim-Zadeh and I. Sorokina, *Mid-Infrared Coherent Sources and Applications* (Springer, 2007).

[71] A. Bogoni, X. X. Wu, Z. Bakhtiari, S. Nuccio, and A. E. Willner, "640 Gbits/s photonic logic gates," Optics Letters **35**, 3955-3957 (2010).

[72] A. Galvanauskas, K. K. Wong, K. El Hadi, M. Hofer, M. E. Fermann, D. Harter, M. H. Chou, and M. M. Fejer, "Amplification in 1.2-1.7 mu m communication window using OPA in PPLN waveguides," Electronics Letters **35**, 731-733 (1999).

[73] S. Barz, G. Cronenberg, A. Zeilinger, and P. Walther, "Heralded generation of entangled photon pairs," Nature Photonics **4**, 553-556 (2010).

[74] A. B. Sugiharto, C. M. Johnson, H. B. De Aguiar, L. Alloatti, and S. Roke, "Generation and application of high power femtosecond pulses in the vibrational fingerprint region," Applied Physics B-lasers and Optics **91**, 315-318 (2008).

[75] S. V. Rao, K. Moutzouris, and M. Ebrahimzadeh, "Nonlinear frequency conversion in semiconductor optical waveguides using birefringent, modal and quasi-phase-matching techniques," Journal of Optics A-pure and Applied Optics **6**, 569-584 (2004).

[76] R. Soref, "Mid-infrared photonics in silicon and germanium," Nature Photonics **4**, 495-497 (2010).

[77] X. P. Liu, J. B. Driscoll, J. I. Dadap, R. M. Osgood, S. Assefa, Y. A. Vlasov, and W. M. J. Green, "Self-phase modulation and nonlinear loss in silicon nanophotonic wires

near the mid-infrared two-photon absorption edge," Optics Express **19**, 7778-7789 (2011).

[78] A. Spott, Y. Liu, T. Baehr-Jones, R. Ilic, and M. Hochberg, "Silicon waveguides and ring resonators at 5.5 mu m," Applied Physics Letters **97**, 213501-213501 (2010).

[79] F. X. Li, S. D. Jackson, C. Grillet, E. Magi, D. Hudson, S. J. Madden, Y. Moghe, C. O'Brien, A. Read, S. G. Duvall, P. Atanackovic, B. J. Eggleton, and D. J. Moss, "Low propagation loss silicon-on-sapphire waveguides for the mid-infrared," Optics Express **19**, 15212-15220 (2011).

[80] S. Zlatanovic, J. S. Park, S. Moro, J. M. C. Boggio, I. B. Divliansky, N. Alic, S. Mookherjea, and S. Radic, "Mid-infrared wavelength conversion in silicon waveguides using ultracompact telecom-band-derived pump source RID D-4869-2011," Nature Photonics **4**, 561-564 (2010).

[81] R. Shankar, R. Leijssen, I. Bulu, and M. Loncar, "Mid-infrared photonic crystal cavities in silicon," Optics Express **19**, 5579-5586 (2011).

[82] V. Raghunathan, D. Borlaug, R. R. Rice, and B. Jalali, "Demonstration of a mid-infrared silicon Raman amplifier," Optics Express **15**, 14355-14362 (2007).

[83] M. M. Milosevic, P. S. Matavulj, P. Y. Y. Yang, A. Bagolini, and G. Z. Mashanovich, "Rib waveguides for mid-infrared silicon photonics," Journal of the Optical Society of America B-optical Physics **26**, 1760-1766 (2009).

[84] M. A. Foster, A. C. Turner, J. E. Sharping, B. S. Schmidt, M. Lipson, and A. L. Gaeta, "Broad-band optical parametric gain on a silicon photonic chip," Nature **441**, 960-963 (2006).

[85] B. Kuyken, H. Ji, S. Clemmen, S. K. Selvaraja, H. Hu, M. Pu, M. Galili, P. Jeppesen, G. Morthier, S. Massar, L. K. Oxenlowe, G. Roelkens, and R. Baets, "Nonlinear properties of and nonlinear processing in hydrogenated amorphous silicon waveguides," Optics Express **19**, B146-153 (2011).

[86] M. Cazzanelli, F. Bianco, E. Borga, G. Pucker, M. Ghulinyan, E. Degoli, E. Luppi, V. Veniard, S. Ossicini, D. Modotto, S. Wabnitz, R. Pierobon, and L. Pavesi, "Second-harmonic generation in silicon waveguides strained by silicon nitride," Nature Materials **11**, 148-154 (2012).

[87] N. K. Hon, K. K. Tsia, D. R. Solli, and B. Jalali, "Periodically poled silicon," Applied Physics Letters **94**, 091116-091116 (2009).

[88] I. Avrutsky and R. Soref, "Phase-matched sum frequency generation in strained silicon waveguides using their second-order nonlinear optical susceptibility," Optics Express **19**, 21707-21716 (2011).

[89] B. Chmielak, M. Waldow, C. Matheisen, C. Ripperda, J. Bolten, T. Wahlbrink, M. Nagel, F. Merget, and H. Kurz, "Pockels effect based fully integrated, strained silicon electro-optic modulator," Optics Express **19**, 17212-17219 (2011).

[90] M. Jazbinsek, L. Mutter, and P. Gunter, "Photonic Applications With the Organic Nonlinear Optical Crystal DAST," Ieee Journal of Selected Topics In Quantum Electronics **14**, 1298-1311 (2008).

[91] M. Zhu, H. Liu, X. Li, N. Huang, Q. Sun, J. Wen, and Z. Wang, "Ultrabroadband flat dispersion tailoring of dual-slot silicon waveguides," Optics Express **20**, 15899-15907 (2012).

[92] C. M. Herzinger, B. Johs, W. A. McGahan, J. A. Woollam, and W. Paulson, "Ellipsometric determination of optical constants for silicon and thermally grown silicon dioxide via a multi-sample, multi-wavelength, multi-angle investigation," Journal Of Applied Physics **83**, 3323-3336 (1998).

[93] E. D. Palik, *Handbook of optical constants of solids* (Academic Press, 1997).

[94] E. Jordana, J. M. Fedeli, P. Lyan, J. P. Colonna, P. Gautier, N. Daldosso, L. Pavesi, Y. Lebour, P. Pellegrino, B. Garrido, J. Blasco, F. Cuesta-Soto, and P. Sanchis, "Deep-UV lithography fabrication of slot waveguides and sandwiched waveguides for Nonlinear applications," 2007 4th Ieee International Conference On Group IV Photonics, 217-219 (2007).

[95] A. D. Bristow, N. Rotenberg, and H. M. van Driel, "Two-photon absorption and Kerr coefficients of silicon for 850-2200 nm," Applied Physics Letters **90**, 191104-191104 (2007).

[96] T. Vallaitis, S. Bogatscher, L. Alloatti, P. Dumon, R. Baets, M. L. Scimeca, I. Biaggio, F. Diederich, C. Koos, W. Freude, and J. Leuthold, "Optical properties of highly nonlinear silicon-organic hybrid (SOH) waveguide geometries," Optics Express **17**, 17357-17368 (2009).

[97] R. Kitamura, L. Pilon, and M. Jonasz, "Optical constants of silica glass from extreme ultraviolet to far infrared at near room temperature," Applied Optics **46**, 8118-8133 (2007).

[98] C. G. Poulton, C. Koos, M. Fujii, A. Pfrang, T. Schimmel, J. Leuthold, and W. Freude, "Radiation modes and roughness loss in high index-contrast waveguides," IEEE Journal of Selected Topics in Quantum Electronics **12**, 1306-1321 (2006).

[99] R. Ding, T. Baehr-Jones, W. J. Kim, X. G. Xiong, R. Bojko, J. M. Fedeli, M. Fournier, and M. Hochberg, "Low-loss strip-loaded slot waveguides in Silicon-on-Insulator," Optics Express **18**, 25061-25067 (2010).

[100] R. Sun, P. Dong, N. N. Feng, C. Y. Hong, J. Michel, M. Lipson, and L. Kimerling, "Horizontal single and multiple slot waveguides: optical transmission at lambda=1550 nm," Optics Express **15**, 17967-17972 (2007).

[101] J. Leuthold, J. Eckner, E. Gamper, P. A. Besse, and H. Melchior, "Multimode interference couplers for the conversion and combining of zero- and first-order modes," Journal of Lightwave Technology **16**, 1228-1239 (1998).

[102] B. Esembeson, M. L. Scimeca, T. Michinobu, F. Diederich, and I. Biaggio, "A High-Optical Quality Supramolecular Assembly for Third-Order Integrated Nonlinear Optics," Advanced Materials **20**, 4584-4587 (2008).

[103] C. Koos, L. Jacome, C. Poulton, J. Leuthold, and W. Freude, "Nonlinear silicon-on-insulator waveguides for all-optical signal processing," Optics Express **15**, 5976-5990 (2007).

[104] C. R. Doerr and L. Chen, "Monolithic PDM-DQPSK receiver in silicon," Proceedings of European Conference on Optical Communication (ECOC), paper PD3_6 (2010).

[105] S. S. Djordjevic, L. W. Luo, S. Ibrahim, N. K. Fontaine, C. B. Poitras, B. Guan, L. Zhou, K. Okamoto, Z. Ding, M. Lipson, and S. J. B. Yoo, "Fully Reconfigurable Silicon Photonic Lattice Filters With Four Cascaded Unit Cells," Photonics Technology Letters, IEEE **23**, 42-44 (2011).

[106] A. Sneh and K. M. Johnson, "High-speed continuously tunable liquid crystal filter for WDM networks," Journal of Lightwave Technology **14**, 1067-1080 (1996).

[107] D. Hillerkuss, M. Winter, M. Teschke, A. Marculescu, J. Li, G. Sigurdsson, K. Worms, S. Ben Ezra, N. Narkiss, W. Freude, and J. Leuthold, "Simple all-optical FFT scheme enabling Tbit/s real-time signal processing," Optics Express **18**, 9324-9340 (2010).

[108] N. Hattasan, W. De Cort, J. Beeckman, K. Neyts, and R. Baets, "Tunable Silicon-on-Insulator based integrated optical filters with liquid crystal cladding," in *LEOS Annual Meeting Conference Proceedings, 2009. LEOS '09. IEEE*, 189-190.

[109] W. D. Cort, J. Beeckman, R. James, F. A. Fernandez, R. Baets, and K. Neyts, "Tuning silicon-on-insulator ring resonators with in-plane switching liquid crystals," J. Opt. Soc. Am. B **28**, 79-85 (2011).

[110] J. Li, S. T. Wu, S. Brugioni, R. Meucci, and S. Faetti, "Infrared refractive indices of liquid crystals," Journal of Applied Physics **97**, 073501-073501 (2005).

[111] H. Desmet, W. Bogaerts, A. Adamski, J. Beeckman, K. Neyts, and R. Baets, "Silicon-on-insulator optical waveguides with liquid crystal cladding for switching and tuning," in *Proceedings of European Conference on Optical Communication*, 430-431.

[112] Y. H. Fan, Y. H. Lin, H. W. Ren, S. Gauza, and S. T. Wu, "Fast-response and scattering-free polymer network liquid crystals for infrared light modulators," Applied Physics Letters **84**, 1233-1235 (2004).

[113] M. Kobayashi, H. Terui, M. Kawachi, and J. Noda, "2x2 Optical-waveguide Matrix Switch Using Nematic Liquid-crystal," Ieee Journal of Quantum Electronics **18**, 1603-1610 (1982).

[114] J. C. Irvin, Bell System Tech. **41**, 387-387 (1962).

[115] D. B. M. Klaassen, "A Unified Mobility Model For Device Simulation .1. Model-equations and Concentration-dependence," Solid-State Electronics **35**, 953-959 (1992).

[116] H. Yu, M. Pantouvaki, J. Van Campenhout, D. Korn, K. Komorowska, P. Dumon, Y. L. Li, P. Verheyen, P. Absil, L. Alloatti, D. Hillerkuss, J. Leuthold, R. Baets, and W. Bogaerts, "Performance tradeoff between lateral and interdigitated doping patterns for high speed carrier-depletion based silicon modulators," Optics Express **20**, 12926-12938 (2012).

[117] S. C. Sun and J. D. Plummer, "Electron-mobility In Inversion and Accumulation Layers On Thermally Oxidized Silicon Surfaces," IEEE Transactions on Electron Devices **27**, 1497-1508 (1980).

[118] C. Lombardi, S. Manzini, A. Saporito, and M. Vanzi, "A Physically Based Mobility Model For Numerical-Simulation Of Nonplanar Devices," IEEE Transactions on Computer-Aided Design Of Integrated Circuits And Systems **7**, 1164-1171 (1988).

[119] L. J. van der Pauw, "A method of measuring specific resistivity and Hall effect of disks of arbitrary shape," Philips Res. Rep. **13**, 1-1 (1958).

[120] C. Bulucea, "Recalculation of Irvin Resistivity Curves For Diffused Layers In Silicon Using Updated Bulk Resistivity Data," Solid-State Electronics **36**, 489-493 (1993).

[121] W. Kern, *Handbook of semiconductor wafer cleaning technology* (Noyes Publications, 1993).

[122] W. Kuzmicz, "Ionization of Impurities In Silicon," Solid-State Electronics **29**, 1223-1227 (1986).

[123] F. Mousty, P. Ostoja, and L. Passari, "Relationship Between Resistivity and Phosphorus Concentration in Silicon," Journal of Applied Physics **45**, 4576-4580 (1974).

[124] Ipkiss, "http://www.ipkiss.org."

[125] M. Iwami, K. Okuno, S. Kamei, T. Ito, and A. Hiraki, "Electronic States of Ni-silicides and Its Relation To Metal-silicide-si Interfaces," Journal of the Electrochemical Society **127**, 1542-1545 (1980).

[126] T. Morimoto, T. Ohguro, H. S. Momose, T. Iinuma, I. Kunishima, K. Suguro, I. Katakabe, H. Nakajima, M. Tsuchiaki, M. Ono, Y. Katsumata, and H. Iwai, "Self-aligned Nickel-mono-silicide Technology for High-speed Deep-submicrometer Logic Cmos Ulsi," IEEE Transactions on Electron Devices **42**, 915-922 (1995).

[127] I. H. Malitson, "Interspecimen Comparison of Refractive Index of Fused Silica," Journal of the Optical Society of America **55**, 1205-1205 (1965).

Acknowledgments

This work would have never been possible without the help of several individuals and organizations, and without the exceptionally favorable environment which I found at the Institute of Photonics and Quantum Electronics (IPQ).

To my main supervisor, Prof. Juerg Leuthold, I am particularly grateful not only for supporting my research in any aspect, for arranging collaborations in the most effective way and for financing every sort of project, but also for the immense source of motivation that he gave to me, and for encouraging every new idea, even the most crazy ones such as cooling a silicon chip down to 40 mK.

I thank Prof. Wolfgang Freude for the uncountable discussions, for sharing the enthusiasm of new results, for his teaching, for developing new ideas, as well for improving my language skills (in German, in English and even Latin). The critical and open discussion atmosphere helped and encouraged several results.

I would like to express my special gratitude to Prof. Christian Koos, for the very frequent brainstorming occasions, for his illuminating suggestions, for his enthusiastic work, and the fun we had in discussing every novel idea.

Several people contributed to my work by granting access to their facilities. In particular, I am grateful to Prof. Uli Lemmer who granted access to his cleanroom and to the e-beam metal evaporator, to Prof. Gernot Goll for granting access to the CFN cleanroom and its high-tech equipment. I thank Prof. Martin Wegener, for granting access to the ALD equipment. I thank Prof. Hilbert v. Löhneysen for granting access to his cryogenic setup. I thank Prof. Georg Weiß for granting access to the ^{3}He-^{4}He refrigerator. My very special thanks go to Dr. Christoph Sürgers for discussing and supporting several low-temperature experiments and for introducing me to the use of several fascinating machines.

I thank Prof. Ingmar Kallfass for his help in designing high-speed silicon transistors. I thank Prof. Thomas Zwick, for the high-speed electrical equipment. I thank Sebastian Diebold, Philipp Pahl, Heiko Gulan and Dr. Mario Pauli for their support in setting up the high-frequency experiments and for several instructive and illuminating discussions about transistors and antennas.

I am highly indebted with Dr. Jean-Marc Fedeli for the time spent in discussing the process flow of the 100 GHz modulator chip, for fabricating the wafer at CEA-LETI, and for giving several technological inputs for solving in-house fabrications issues.

I thank Dr. Pieter Dumon, who coordinated the fabrication of several wafer at Imec, and supported my work in several different occasions. I thank Dr. Shankar Kumar Sevaraja, for fabricating and sharing several interesting chips, as well for the many interesting discussions.

For my "silicon photonics education" I am very grateful to Prof. Lorenzo Pavesi who organized the excellent 5th Optoelectronics Winter School in Trento. I thank Dr. Cary Gunn for the insightful and motivating discussions during this period or in later more official occasions.

I thank Dr. Raluca Dinu and Dr. Baoquen Chen of GigOptix for providing large quantities of reliable nonlinear polymers, and for the help in optimizing the appropriate poling procedure. I thank Prof. Alan Mickelson for receiving me in his laboratories, for his support and the interesting discussions. I thank Xi Chen for the support, even at late hours, during my stay in Colorado.

I thank Dietmar Korn, Dr. Sebastian Köber, Matthias Lauermann, Sascha Mühlbrandt, Jingshi Li, Robert Palmer, Philipp Schindler and all my colleagues for their support and collaboration. I thank my students Julian Gross, Siegwart Bogatscher, Robert Palmer, Claudius Weimann, Christoph Reustle and Christoph Stang for their enthusiasm and productivity. I thank Andreas Frölich for growing many ALD samples and for his ideas.

I thank Mr. Bürger, Mr. Hirsch and Mr. Höhne for building so many mechanical devices and in such short times in the wonderful workshop of the Institute. I thank Oswald Speck, Silvia Diewald, Sebastian Struck, Martin Winkeler and Florian Rupp for the great technical support. I thank our secretaries Ms. Olbrich, Ms. Riemensperger and most specially Bernadette Lehmann for their work and friendliness.

List of Publications

Journal Papers

[J11] **L. Alloatti** et al., "Low-temperature sheet resistance of accumulation/inversion layers in silicon,"In preparation.

[J10] **L. Alloatti**, M. Lauermann, C. Suergers, C. Koos, W. Freude, and J. Leuthold, " Optical absorption in silicon layers in the presence of charge inversion/accumulation or ion implantation" Accepted for publication in APL.

[J9] **L. Alloatti**, S. Diebold, P. Pahl, M. Pauli, J. Li, B. Chen, R. Dinu, M. Fournier, J. Fedeli, T. Zwick, C. Koos, W. Freude, and J. Leuthold, "Over 100 GHz Pockels effect modulator in silicon hybrid technology" Submitted for publication.

[J8] **L. Alloatti**, D. Korn, C. Weimann, C. Koos, W. Freude, and J. Leuthold, "Second-order nonlinear silicon-organic hybrid waveguides," Optics Express **20**, 20506-20515 (2012).

[J7] **L. Alloatti**, D. Korn, R. Palmer, D. Hillerkuss, J. Li, A. Barklund, R. Dinu, J. Wieland, M. Fournier, J. Fedeli, H. Yu, W. Bogaerts, P. Dumon, R. Baets, C. Koos, W. Freude, and J. Leuthold, "42.7 Gbit/s electro-optic modulator in silicon technology," Optics Express **19**, 11841-11851 (2011).

[J6] **L. Alloatti**, "Anomalies and Anderson localization: pi-coupling of the energy bands," Journal of Physics-condensed Matter **21**, 045503-045503 (2009).

[J5] J. Pfeifle, **L. Alloatti**, W. Freude, J. Leuthold, and C. Koos, "Silicon-organic hybrid phase shifter based on a slot waveguide with a liquid-crystal cladding," Optics Express **20**, 15359-15376 (2012).

[J4]. H. Yu, M. Pantouvaki, J. Van Campenhout, D. Korn, K. Komorowska, P. Dumon, Y. L. Li, P. Verheyen, P. Absil, **L. Alloatti**, D. Hillerkuss, J. Leuthold, R. Baets, and W. Bogaerts, "Performance tradeoff between lateral and interdigitated doping patterns for high speed carrier-depletion based silicon modulators," Optics Express **20**, 12926-12938 (2012).

[J3] T. Alasaarela, D. Korn, **L. Alloatti**, A. Saynatjoki, A. Tervonen, R. Palmer, J. Leuthold, W. Freude, and S. Honkanen, "Reduced propagation loss in silicon strip and slot waveguides coated by atomic layer deposition," Optics Express **19**, 11529-11538 (2011).

[J2] T. Vallaitis, S. Bogatscher, **L. Alloatti**, P. Dumon, R. Baets, M. L. Scimeca, I. Biaggio, F. Diederich, C. Koos, W. Freude, and J. Leuthold, "Optical properties of highly nonlinear silicon-organic hybrid (SOH) waveguide geometries," Optics Express **17**, 17357-17368 (2009).

[J1] A. B. Sugiharto, C. M. Johnson, H. B. De Aguiar, **L. Alloatti**, and S. Roke, "Generation and application of high power femtosecond pulses in the vibrational fingerprint region," Applied Physics B-lasers and Optics **91**, 315-318 (2008).

Patents

[P3] **L. Alloatti**, J. Leuthold, W. Freude, C. Koos, D. Korn, and C. Weimann, "Wellenleiter-Bauteil fuer nichtlinear-optische Prozesse zweiter Ordnung," 102012016328.2 (2012).

[P2] **L. Alloatti**, J. Leuthold, W. Freude, C. Koos, D. Korn, and R. Palmer, "Electro-optical device and method for processing an optical signal, USA," U.S.A. 13/460,395 (2012).

[P1] **L. Alloatti**, J. Leuthold, W. Freude, C. Koos, D. Korn, and R. Palmer, "Electro-optical device and method for processing an optical signal, EU," EP 11003562.3-1228 (2012).

Conference Contributions

[C23] **L. Alloatti** et al., "Chi(2) parametric amplifiers, high-speed electro-optic modulators, ultra-low power phase-shifters in silicon-organic hybrid technology," SPIE Photonics West, San Francisco (2013) **[invited]**.

[C22] **L. Alloatti**, J. Pfeifle, J. Mendez, W. Freude, L. J, and C. Koos, "Liquid crystal phase shifter on the SOH platform with ultra-low power consumption," Optical Fiber Communication Conference (OFC'12), Paper OTu1I.5 (2012).

[C21] **L. Alloatti**, D. Korn, D. Hillerkuss, T. Vallaitis, J. Li, R. Bonk, R. Palmer, T. Schellinger, A. Barklund, R. Dinu, J. Wieland, M. Fournier, J. Fedeli, W. Bogaerts, P. Dumon, R. Baets, C. Koos, W. Freude, and J. Leuthold, "Silicon High-Speed Electro-Optic Modulator," Proceedings of IEEE Conference on Group IV Photonics, paper ThC2 (2010).

[C20] **L. Alloatti**, D. Korn, D. Hillerkuss, T. Vallaitis, J. Li, R. Bonk, R. Palmer, T. Schellinger, A. Barklund, R. Dinu, J. Wieland, M. Fournier, J. Fedeli, P. Dumon, R. Baets, C. Koos, W. Freude, and J. Leuthold, "40 Gbit/s Silicon-Organic Hybrid (SOH) Phase Modulator," European Conference on Optical Communication (ECOC 2010), Paper Tu.5.C.4 (2010).

[C19] D. Korn, **L. Alloatti**, M. Lauermann, J. Pfeifle, R. Palmer, P. C. Schindler, W. Freude, C. Koos, J. Leuthold, H. Yu, W. Bogaerts, K. Komorowska, R. Baets, J. Van Campenhout, P. Verheyen, J. Wouters, M. Moelants, P. Absil, M. Dispenza, A. Secchi, M. Jazbinsek, P. Gunter, S. Wehrli, M. Bossard, P. Zakynthinos, and I. Tomkos, "Silicon organic hybrid fabrication platform for integrated circuits," 14th Intern. Conf. on Transparent Optical Networks (ICTON'12) (2012).

[C18] R. Palmer, **L. Alloatti**, D. Korn, W. Heni, P. Schindler, J. Bolten, M. Karl, M. Waldow, T. Wahlbrink, W. Freude, C. Koos, and J. Leuthold, "Highly Efficient Strip-to-Slot Mode Converters," CLEO 2012, Paper CM4M1 (2012).

[C17] D. Korn, H. Yu, D. Hillerkuss, **L. Alloatti**, C. Mattern, K. Komorowska, W. Bogaerts, R. Baets, J. Van Campenhout, P. Verheyen, J. Wouters, M. Moelants, P. Absil, C. Koos, W. Freude, and J. Leuthold, "Detection or modulation at 35 Gbit/s with a standard CMOS-processed optical waveguide," Conf. on Lasers and Electro-Optics (CLEO'12), Paper CTu1A.1 (2012).

[C16] D. Korn, M. Lauermann, P. Appel, **L. Alloatti**, R. Palmer, W. Freude, J. Leuthold, and C. Koos, "First silicon-organic hybrid laser at telecommunication wavelengths," Conf. on Lasers and Electro-Optics (CLEO'12), Paper CTu2I.1 (2012).

[C15] C. Koos, J. Leuthold, W. Freude, **L. Alloatti**, D. Korn, N. Lindenmann, R. Palmer, J. Pfeifle, D. Hillerkuss, and R. Schmogrow, "Silicon Nanophotonics and Photonic Wire Bonding: Technologies for Terabit/s Interconnects," 6th ITG Workshop on Photonic Integration, Laser Optics Berlin (2012).

[C14] H. Yu, W. Bogaerts, K. Komorowska, R. Baets, D. Korn, **L. Alloatti**, D. Hillerkuss, C. Koos, W. Freude, J. Leuthold, J. Campenhout, P. Verheyen, J. Wouters, M. Moelants, and P. Absil, "Doping geometries for 40G carrier-depletion-based silicon optical modulators," Optical Fiber Communication Conference (OFC'12), Paper OW4F.4 (2012).

[C13] W. Freude, **L. Alloatti**, A. Melikyan, R. Palmer, D. Korn, N. Lindenmann, T. Vallaitis, D. Hillerkuss, J. Li, A. Barklund, R. Dinu, J. Wieland, M. Fournier, J. Fedeli, S. Walheim, P. M. Leufke, S. Ulrich, J. Ye, P. Vincze, H. Hahn, H. Yu, W. Bogaerts, P. Dumont, R. Baets, B. Breiten, F. Diederich, M. T. Beels, I. Biaggio, T. Schimmel, C. Koos, and J. Leuthold, "Nonlinear optics on the silicon platform," Optical Fiber Communication Conference (OFC'12), Paper OTh3H.6 (2012).

[C12] P. Schindler, H. Fischer, H. Haisch, N. Lindenmann, **L. Alloatti**, R. Palmer, D. Korn, D. Hillerkuss, R. Schmogrow, A. Gerster, W. Freude, J. Leuthold, and C. Koos, "Optical Technologies for Off-Chip and On-Chip Interconnects," DesignCon 2012, Santa Clara (CA) (2012).

[C11] A. Säynätjoki, L. Karvonen, T. Alasaarela, D. Korn, **L. Alloatti**, A. Tervonen, R. Palmer, C. Koos, J. Leuthold, W. Freude, and S. K. Honkanen, "Loss reduction of silicon slot waveguides with ALD-grown thin films," OPTO SPIE Photonics West - Society of Photo-Optical Instrumentation Engineers (OPTO-SPIE'12), Paper 8266 (2012).

[C10] C. Koos, **L. Alloatti**, D. Korn, R. Palmer, T. Vallaitis, R. Bonk, D. Hillerkuss, J. Li, W. Bogaerts, P. Dumon, R. Baetes, M. L. Scimeca, I. Biaggio, A. Barklund, R. Dinu, J. Wieland, M. Fournier, J. Fedeli, W. Freude, and J. Leuthold, "Silicon nanophotonics and silicon-organic hybrid (SOH) integration," Gerneral Assembly and Scientific Symposium, 2011 XXXth URSI Istanbul (2011).

[C9] J. Leuthold, W. Freude, C. Koos, **L. Alloatti**, D. Korn, R. Palmer, and J. M. Brosi, "Slotted Photonic Crystal Slow Light Modulators," Integrated Photonics Research, Silicon and Nanophotonics (IPRSN), Paper JTuA3 (2011).

[C8] C. Koos, **L. Alloatti**, D. Korn, R. Palmer, D. Hillerkuss, J. Li, A. Barklund, R. Dinu, J. Wieland, M. Fournier, J. Fedeli, H. Yu, W. Bogaerts, P. Dumon, R. Baets, W. Freude, and L. J, "Silicon-Organic Hybrid (SOH) Electro-Optical Devices," Integrated Photonics Research, Silicon and Nano Photonics (IPR) Topical Meeting of the OSA, Paper IWF1 (2011).

[C7] R. Palmer, **L. Alloatti**, D. Korn, M. Moosmann, K. Huska, U. Lemmer, D. Gerthsen, T. Schimmel, W. Freude, K. C, and L. J, "Smooth and ultra-precise silicon nanowires fabricated by conventional optical lithography," CLEO 2011, Paper CThZ1 (2011).

[C6] W. Freude, **L. Alloatti**, T. Vallaitis, D. Korn, D. Hillerkuss, R. Bonk, R. Palmer, J. Li, T. Schellinger, M. Fournier, J. Fedeli, W. Bogaerts, P. Dumon, R. Baets, A. Barklund, R. Dinu, J. Wieland, and M. L. Scimeca, "High-speed signal processing with silicon-organic hybrid devices," European Optical Society Annual Meeting (EOS'10) (2010).

[C5] J. Leuthold, C. Koos, W. Freude, T. Vallaitis, **L. Alloatti**, D. Korn, P. Dumon, W. Bogaerts, R. Baets, I. Biaggio, and F. Diederich, "All-optical signal processing with silicon-organic hybrid slot waveguides," OSA Integrated Photonics Research, Silicon and Nano Photonics (IPR), Paper IWC1 (2010).

[C4] W. Freude, J. Leuthold, **L. Alloatti**, T. Vallaitis, D. Korn, R. Palmer, C. Koos, J. M. Brosi, P. Dumon, R. Baets, M. L. Scimeca, I. Biaggio, B. Breiten, F. Diederich, A. Barklund, R. Dinu, and J. Wieland, "100 Gbit/s electro-optic modulator and 56 Gbit/s wavelength converter for DQPSK data in silicon-organic hybrid (SOH) technology," Proceedings of IEEE Conference on Novel Waveguiding, Structures and Phenomena, Summer Topicals, Paper WB2.1 (2010).

[C3] W. Freude, T. Vallaitis, C. Koos, J. M. Brosi, **L. Alloatti**, P. Dumon, R. Baets, M. L. Scimeca, I. Biaggio, B. Breiten, F. Diederich, and J. Leuthold, "Ultrafast silicon-organic hybrid (SOH) photonics," Proc. Conf. on Lasers and Electro-Optics (CLEO/IQEC'10), Paper CThR1 (2010).

[C2] W. Freude, C. Koos, T. Vallaitis, J. M. Brosi, **L. Alloatti**, S. Bogatscher, P. Dumon, R. Baets, M. L. Scimeca, I. Biaggio, F. Diederich, and J. Leuthold, "Silicon-organic hybrid (SOH) photonics for ultra-fast optical signal processing," The 22nd Annual Meeting of the IEEE Photonics Society (LEOS 2009), Paper WI 2 (2009).

[C1] J. Leuthold, W. Freude, and **L. Alloatti**, "100 Gb/s Silicon Organic Hybrid (SOH) Devices," Workshop on Tunable and Active Silicon Photonics, Hamburg (2008).

Curriculum Vitae

Luca Alloatti

Born on July 23rd, 1980, in Turin, Italy

Citizenships: Italian and Swiss

Education:

1999-2004	Studies of Physics at the *Scuola Normale Superiore di Pisa*, Italy Degree: "*Diploma di Licenza*" Final Mark: 70/70 *cum laude*
1999-2004 (in parallel)	Master Studies of Physics at the University of Pisa Thesis: Conductance in almost one-dimensional quantum systems Degree: Dipl. Phys. Final Mark: 110/110 *cum laude*
1994-1999	Secondary Scool (*Liceo Scientifico*) in Aosta, Italy

Working Experience:

2008-2012	PhD studies at the Karlsruhe Institute of Technology (KIT), Institute of Photonics and Quantum Electronics (IPQ), Karlsruhe, Germany
2007-2008	Employee at the Max Planck Institute for Metal Research, Stuttgart, Germany Reserch Topic: mid-IR and far-IR femtosecond nonlinear spectroscopy by sum-frequency generation
2004-2006	Employee at the Max Planck Institute for Metal Research, Stuttgart, Germany Reserch Topic: Statistical Mechanics

Scholarships:

2010	"Travel Scholarship" for visiting the University of Colorado, Boulder, USA and the company Lumera, GigOptix, Seattle, USA
1999-2004	Scholarship from *the Scuola Normale Superiore di Pisa* (accomodation, meals, university fees, pocket-money)

Honors and Awards:

2009	"Most active student award" at the 5th Optoelectronics and Photonics Winter School, Trento, Italy
2006	"*Diploma di Licenza*" from the *Scuola Normale Superiore di Pisa*, Italy
1999	Prize for the best experimental work at the Italian *national* competition of physics

Language Skills:

Italian	Native language
French	Bilingual
English	Fluent
German	Fluent
Swiss-German	Intermediate

Karlsruhe Series in Photonics & Communications
KIT, Institute of Photonics and Quantum Electronics (IPQ)
(ISSN 1865-1100)

Die Bände sind unter www.ksp.kit.edu als PDF frei verfügbar
oder als Druckausgabe bestellbar.

Band 1 Christian Koos
Nanophotonic Devices for Linear and Nonlinear Optical Signal Processing. 2008
ISBN 978-3-86644-178-1

Band 2 Ayan Maitra
Nonlinear Resonators for All-Optical Signal Processing. 2009
ISBN 978-3-86644-150-7

Band 3 Jin Wang
Pattern Effect Mitigation Techniques for All-Optical Wavelength Converters Based on Semiconductor Optical Amplifiers. 2008
ISBN 978-3-86644-276-4

Band 4 Jan-Michael Brosi
Slow-Light Photonic Crystal Devices for High-Speed Optical Signal Processing. 2009
ISBN 978-3-86644-313-6

Band 5 Christoph Dyroff
Tunable Diode-Laser Absorption Spectroscopy for Trace-Gas Measurements with High Sensitivity and Low Drift. 2009
ISBN 978-3-86644-328-0

Band 6 Philipp Vorreau
An Optical Grooming Switch for High-Speed Traffic Aggregation in Time, Space and Wavelength. 2010
ISBN 978-3-86644-502-4

Band 7 Thomas Vallaitis
Ultrafast Nonlinear Silicon Waveguides and Quantum Dot Semiconductor Optical Amplifiers. 2011
ISBN 978-3-86644-748-6

Band 8 René Bonk
Linear and Nonlinear Semiconductor Optical Amplifiers for Next-Generation Optical Networks. 2013
ISBN 978-3-86644-956-5

Karlsruhe Series in Photonics & Communications
KIT, Institute of Photonics and Quantum Electronics (IPQ)
(ISSN 1865-1100)

Band 9 David Hillerkuss
Single-Laser Multi-Terabit/s Systems. 2013
ISBN 978-3-86644-991-6

Band 10 Moritz Röger
Optically Powered Highly Energy-efficient Sensor Networks. 2013
ISBN 978-3-86644-972-5

Band 11 Jingshi Li
Optical Delay Interferometers and their Application for Self-coherent Detection. 2013
ISBN 978-3-86644-957-2

Band 12 Luca Alloatti
High-Speed, Low-Power and Mid-IR Silicon Photonics Applications. 2013
ISBN 978-3-7315-0056-8